Frank Fleischer

Analysis and Fitting of Random Tessellation Models

Frank Fleischer

Analysis and Fitting of Random Tessellation Models

Applications in Telecommunication and Cell Biology

VDM Verlag Dr. Müller

Imprint

Bibliographic information by the German National Library: The German National Library lists this publication at the German National Bibliography; detailed bibliographic information is available on the Internet at http://dnb.d-nb.de.

Cover image: www.purestockx.com

Publisher:
VDM Verlag Dr. Müller Aktiengesellschaft & Co. KG , Dudweiler Landstr. 125 a, 66123 Saarbrücken, Germany,
Phone +49 681 9100-698, Fax +49 681 9100-988,
Email: info@vdm-verlag.de

Zugl.: Ulm, Universität, Diss., 2007

Produced in USA and UK by:
Lightning Source Inc., La Vergne, Tennessee, USA
Lightning Source UK Ltd., Milton Keynes, UK
BookSurge LLC, 5341 Dorchester Road, Suite 16, North Charleston, SC 29418, USA

ISBN: 978-3-8364-8743-6

Contents

Chapter 1

Introduction

1.1 Motivation

This thesis has been motivated by two ongoing research projects of the Institute of Stochastics at Ulm University. In the first project, which is performed in cooperation with France Télécom R&D, Paris, it is dealt with the modelling and the analysis of network structures that occur in the field of telecommunication. In particular street systems, for example of urban regions like Paris, but also telecommunication networks on nationwide scales are analysed with respect to their geometrical structures and cost characteristics connected to them. Figure 1.1 shows such an infrastructure system for the example of Paris, while in Figure 1.2 a magnified cutout is displayed, where the actual measurements are shown in red and the polygonal connections between them in black. Figure 1.3a illustrates a modelling approach for telecommunication networks in an urban area. Here, the streets are represented by random lines. Different types of telecommunication equipment are then placed on these lines according to a given random process. In Figure 1.3b we see a realisation for a modelling approach with respect to telecommunication networks on a nationwide scale. In this case the network equipment is placed randomly in the plane according to a given random process that reflects the differences between urban and rural districts by assigning different intensities. The subscribers are also distributed randomly according to a similar mechanism, thereby reflecting the underlying population distribution. Aims of such modelling approaches are, for example, to obtain inference about connected costs like the mean distance from a subscriber to its nearest telecommunication equipment located along the streets or about the capacities needed to connect all the subscribers.

In the second project which is performed jointly with co-workers from the Institute of Internal Medicine 1 at Ulm University, the Central Electron Microscopy Facility at

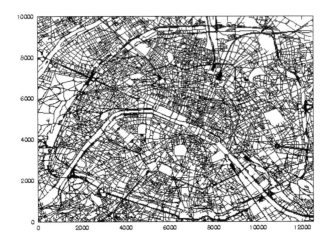

Figure 1.1: Urban infrastructure of Paris

Figure 1.2: Infrastructure in a region of Paris

a) *Realisation of a mathematical model that describes the serving zones of a telecommunication network in an urban area*

b) *Realisation of a mathematical model that describes the population density and the serving zones of a telecommunication network on a nationwide scale*

Figure 1.3: Modelling of telecommunication networks in urban areas and on a nationwide scale

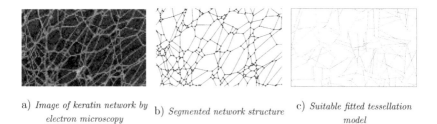

a) *Image of keratin network by*   b) *Segmented network structure*   c) *Suitable fitted tessellation*
   electron microscopy   *model*

Figure 1.4: Investigation of keratin filament structures

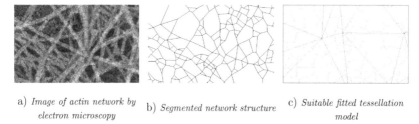

a) *Image of actin network by*   b) *Segmented network structure*   c) *Suitable fitted tessellation*
   electron microscopy   *model*

Figure 1.5: Investigation of actin filament structures

Ulm University, the Laboratory of Cell and Computational Biology at the University of California at Davis, the Department of Physics at the University of Leipzig, and the Department of Biology at the University of Pennsylvania, intracellular structures in human cells are analysed. In particular, network structures formed by the cytoskeleton are investigated with respect to their, both geometric and random, nature. Figure 1.4 shows the first data set investigated that consists of images from electron microscopy showing keratin filament structures in the cytoskeleton. They have been segmented using tools of morphological image analysis and afterwards modelled by a suitable tessellation (network) structure. In Figure 1.5 a second type of filamentous network structure in the cytoskeleton constructed by actin fibres is displayed which also has been segmented and modelled by a suitable tessellation model.

Although at first sight these two projects do not seem to have too much in common it became more and more obvious during the course of these projects that the mathematical models, tools and techniques involved are quite similar. The data sets that are analysed represent network structures in both cases, either on a macroscopic (telecommunication data) or on a microscopic, even nanoscopic, scale (cytoskeletal data). The

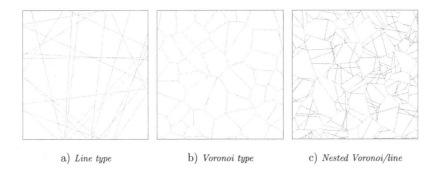

a) *Line type*   b) *Voronoi type*   c) *Nested Voronoi/line*

Figure 1.6: Different tessellation models

connections between two vertices or nodes of these networks can be approximated quite well by linear segments leading to almost polygon–like structures, where in some cases of course convexity can not be guaranteed. This almost polygon–like structures have led in both projects to an approach of modelling the network structures by so–called random tessellations or in other words random mosaics that can be considered as a random partition of the space into non–empty, non–overlapping polygons. Examples for such tessellation models are shown in Figure 1.6, where in Figure 1.6a the tessellation is formed by random lines. In Figure 1.6b a so–called Voronoi tessellation is displayed which is based on a nearest neighbour principle, whereas in Figure 1.6c a combination of the first two tessellations based on a nesting of the Voronoi tessellation into the line tessellation is shown. A need for the determination of appropriate tessellation models that are suitable to represent the observed networks in a proper way while on the other hand are still remaining mathematically tractable by theoretical formulae or at least by simulation studies is therefore a key necessity. Additionally, after the determination of such a tessellation model it is often of interest to compute certain model characteristics. Such characteristics comprise for example the mean perimeter of the polygons forming the tessellation, the mean distance of a randomly chosen point to the nuclei of the polygon it is located in, or the mean diameter of the maximal incircle of the polygon, sometimes also called the average mesh size.

More specifically, with respect to the first project that is located in the field of telecommunication a main goal that is achieved in this thesis is the derivation and application of efficient estimators for cost functionals in two–level hierarchical models that are based on random point processes and random tessellation models. The results of such estimations can then later be used for example to perform cost calculations or to do risk analysis with respect to tail distributions or in other words extreme values occuring. In order to derive such efficient estimators, of course, some basic knowledge about topics of stochastic geometry is necessary. It is also necessary to develop simulation algorithms for the typical cell of specific types of random tessellations that are involved in the two–level hierarchical models. Typical cell in this context means that from all cells occuring, a representant is chosen according to the distribution that reflects the probability of occurrence.

With regard to the second project that is based on image data of filament networks in the cytoskeleton methods are described how to segment these images in order to obtain graph structures that are suitable for later analysis. The graph structures are then used in order to perform model fitting with random tessellation models where the fitting algorithm is based on the comparison of estimated global characteristics like the number of vertices or the total length of the edges to theoretical characteristics for the models that can be computed by mean value formulae. In particular, for the example dealing with actin networks, it is explained how to estimate a specific characteristic, the average mesh size, based on the fitted tessellation model. This mean mesh size is afterwards useful in order to compute characteristics for cell elasticity that determines the ability of the cell to move and migrate.

This thesis will show some of the results that were obtained in these two research projects and will thereby try to convince the reader of the universality of the applied techniques of stochastic geometry and morphological image analysis with respect to their applicability in various fields of scientific and industrial research. Basically this means that methods that are applied in order to solve a specific type of problem for one field of application can often be slightly modified and used to solve a problem in a field of application that is quite different with respect to its scale and other types of characteristics connected to it. To summarize things, the purposes of this thesis are

- to develop and describe efficient algorithms for the simulation of the typical cell for different types of tessellation models. Additionally to the fact that valuable inference is obtained with respect to characteristics of the typical cell itself, these simulation algorithms serve as a cornerstone for an efficient cost analysis;

- to perform such an efficient cost analysis based on two–level hierarchical models. Such a cost analysis can, for example, be used to allow for realistic cost calculations in the sector of telecommunication or to obtain information about biological processes connected with the transport of vesicles in cell membrane trafficking;

- to generally discuss some possible software tests with respect to implementations that occur, with a special regard to the fact that implementations in this context often have a random input or output;

- to show some useful techniques and methods from morphological image analysis in order to preprocess given image data such that a later statistical analysis is made possible;

- to perform such a statistical analysis for two examples from cell biology where samples from the cytoskeleton are investigated. Here, apart from a basic statistical examination, especially a model fitting algorithm for random tessellation models is of interest. Such a model fitting approach is of course not restricted to examples on a microscopic scale but can be applied, for example, to urban infrastructure data;

- to show by all the points given above that the techniques and tools from stochastic geometry and image analysis that have been applied are quite universal in the sense that they can be used in various fields with problems that have a scale ranging from macroscopic to microscopic or even nanoscopic.

1.2 Outline

After the introduction found in this chapter, some basic concepts of stochastic geometry are given in Chapter 2. The aim is to provide the reader with enough foundations of this particular mathematical discipline in order to be able to introduce the notion of random tessellation, the one object that is common to all applications discussed in later chapters. In Sections 2.1 random closed sets are defined, while in Section 2.2 random point processes are introduced, both for the unmarked as well as for the marked case. In this section also Palm distributions of random point processes and Neveu's exchange formula with respect to such Palm distributions are given that will become important in Chapters 3 and 4. Other important models from stochastic geometry that will play a role in Chapters 3 and 4 are the Boolean model and modulated Poisson point processes that are based on the Boolean model. They are therefore introduced in Section 2.3. In Sections 2.4, finally, deterministic as well as random tessellations are defined. Various examples are provided that will become important during the course of this thesis like Poisson–Voronoi tessellations, Poisson line tessellations, Poisson–Delaunay tessellations, Cox–Voronoi tessellations, modulated tessellations, superpositions and nestings. Note that in this thesis mostly the planar case is discussed but that there often exists a canonical extension of these topics to higher dimensions.

In Chapter 3 some efficient algorithms for the simulation of the typical cell for various tessellation models are introduced. Apart from indicating interesting results for characteristics of the typical cell on its own right, we will apply these techniques in Chapter 4 in order to derive efficient estimators for cost functionals in two–level hierarchical models. In Section 3.1 some general aspects of such algorithms for the simulation of the typical cell for a random tessellation are discussed. We start with algorithms for the simulation of Poisson point processes which provides a basis for all simulation algorithms given in this chapter. Slivnyak's theorem is introduced that allows to express the Palm distribution of a Poisson point process by its (unconditional) distribution and by adding a point in the origin. Based on the simulation algorithm for the Poisson point process and on Slivnyak's theorem a simulation algorithm for the typical cell of a Poisson–Voronoi tessellation is provided. This simulation algorithm is extended in Section 3.2 in order to simulate the typical cell of a Cox–Voronoi tessellation, where the Cox–Voronoi tessellation is induced by a stationary Cox point processes whose random driving measure is concentrated on the lines of a Poisson line tessellation. Some numerical results for characteristics of the typical cell like empirical distributions for the area and the perimeter are given. In Section 3.3 an algorithm for the simulation of the typical cell for a different type of tessellation, the modulated Poisson–Voronoi tessellation, is introduced. Here we have the case that the Voronoi tessellation is based on a Cox point process whose random driving measure is induced by a Boolean model with circular grains of a fixed or at least bounded radius. Numerical results are provided for some specific parameter configurations with respect to characteristics like the distribution of the area of the typical cell and the number of vertices.

The aim of Chapter 4 is to derive efficient estimators for certain cost functionals in two different two–level hierarchical models. As a preliminary step some basic notions of graph theory are discussed in Section 4.1. In particular, after the definition of a graph some well–known algorithms for the computation of shortest paths and their associated lengths in a given graph like Dijkstra's algorithm are introduced.

The first two–level hierarchical model that is discussed in Section 4.2 is based on two Cox point processes with random driving measures that are concentrated on the lines of a Poisson line process. A specific characteristic of interest is the shortest path length, i.e., the distance along the lines between a point of the lower–level point process and its nearest neighbour (in the Euclidean sense) belonging to the higher–level point process. A characteristic that is closely related to the shortest path length in this two–level hierarchical model is the subscriber line length. Here, instead of locating the points of the lower–level point process on the lines of the Poisson line process, they are located randomly in the plane and projected afterwards to the nearest point on the line system that still belongs to the same influence zone, i.e., that has to be located in the Voronoi cell of the nearest point belonging to the higher level point process. The subscriber line length is then the distance along the lines of the Poisson line process from the projected point to the nuclei of the Voronoi cell (the nearest neighbouring point that belongs to the higher–level point process). Efficient estimators for mean values of both the shortest path length as well as the mean subscriber line length are derived that are based on the algorithms for the simulation of the typical cell introduced in Chapter 3 and on Neveu's exchange formula for Palm distributions described in Section 2.2. Some results of Monte–Carlo simulations are provided that in conjunction with a scaling invariance property of the model enlight some possibilities for a usage of these techniques with respect to cost calculation.

In Section 4.3 another two–level hierarchical model is discussed that is based on modulated Poisson point processes introduced in Section 2.3. In particular the mean distance from a point of the lower–level point process to its nearest neighbour belonging to the higher–level point process is investigated. For this purpose an efficient estimator is derived that is based on the algorithms for the simulation of the typical cell introduced in Chapter 3 and on Neveu's exchange formula for Palm distributions described in Section 2.2 as it has been the case for the shortest path length and the mean subscriber line length. Some numerical examples are given and it is discussed how to perform an efficient cost analysis by the usage of a scaling invariance effect.

In Chapter 5 some methods for software tests with respect to implementations that have a random input or output are discussed. In this context tests that are based on a statistical oracle are a key concept which is introduced in Section 5.2. Some examples for implementations of the algorithms described in Chapters 3 and 4 are provided. In Section 5.3 the statistical oracle is combined with a technique called metamorphic testing in order to obtain a second class of software tests for implementations with random input or output. Also, for this class of tests, examples are given based on implementations of simulation algorithms for the typical cell of different tessellation models. A third class of tests is derived in Section 5.4 by combining a statistical oracle, a metamorphic relation and another given (and already tested) implementation, the so–called gold standard. After an introduction of this testing technique some examples for applications of this class of tests are provided. The chapter ends with a summary of the testing methods and a comparison between them.

The topic of Chapter 6 is the introduction of some concepts of morphological image analysis. They will be used in Chapter 8 to preprocess given sample images from electron microscopy in order to allow for a subsequent statistical analysis and fitting of suitable random tessellation models. In Section 6.1 digital grids and different types of digital images like grey scale or RGB–images are defined. Additionally, methods of image filtering are discussed with a special regard to low–pass filters like average or Gaussian filters. Section 6.2 is dedicated to the explanation of a specific algorithm for morphological image analysis, the skeletonization by morphological operators which reduces a given structure to a new structure with line width one that still resembles features including the number of connected components in the old structure. A method that is closely related to skeletonization is morphological watershed transformation which is discussed in Section 6.3. In particular the algorithm of watershed transformation by immersion is explained in detail. In the last section of this chapter further morphological operations are described that are applied in Chapter 8 in order to enhance the results. Such operations comprise, for example, an iterative pruning and merging of the obtained binary structures.

In Chapter 7 a method for the fitting of random tessellation models to given network structures is introduced. This method is applied in Chapter 8 to examples from cell biology but due to its universality it can be applied to other types of network structures like urban infrastructure data (cmp. [31], [92]). In Section 7.1 the used characteristics of the input data as well as unbiased estimators for these characteristics are discussed. The fitting procedure itself together with the choice of the distance function and the optimal model is explained in Section 7.2. The pool of possible tessellation models considered in later applications consists of Poisson–type basic tessellation models and either one–fold nestings or one–fold superpositions.

Chapter 8 shows two applications from cell biology for the methods described in Chapters 6 and 7. In the first application, discussed in Section 8.1, a statistical analysis of keratin filament structures is performed. Such keratin filament structures are found in the cytoskeleton of epithelia cancer cells and play an important role in cell mechanisms like motility and structural integrity. The main goal of this application is to detect and describe morphological changes in the keratin network architecture that are caused by the injection of substances that enhance tumor growth. After the description of image acquisition and segmentation a basic statistical analysis of the filaments by means of investigating their orientations and lengths is performed. One–fold nested random tessellation models are fitted to the network structures using the fitting procedure described in Chapter 7. The results of this fitting procedure show that indeed there is a restructuring of the architecture in the keratin network caused by the addition of tumor–enhancing substances and that it can be qualitatively as well as as quantitatively described by the mathematical techniques described in this thesis.

In the second example from cell biology, introduced in Section 8.2, actin filament networks are investigated that are, as it is the case for the keratin filaments, located in the cytoskeleton of different types of cells and that are responsible for the regulation of the elasiticity of whole cells, thereby influencing cell migration. The aim of this analysis is to determine a random tessellation that is capable of reflecting the basic characteristics of the network structures given in sample images from electron microscopy. Afterwards, based on these tessellation models, approximations for characteristics like the elastic shear modulus are computed that are measuring cell elasticity. In Section 8.2.1 the applied image segmentation algorithm is described in detail which is based on the morphological watershed transformation given in Section 6.3. The results for the fitting procedure of a one–fold superposed random tessellation model are documented in Section 8.2.2 leading to an optimal tessellation model that is used in Section 8.2.3 in order to derive estimates for the elastic shear modulus of the cell. These estimations are then compared to an estimation that is purely based on the actin concentration. Therefore, this application demonstrates that the approach of fitting random tessellation models is capable of providing useful information about various kinds of characteristics for biological cells.

In Chapter 9 the results of this thesis are summarized and an outlook to further interesting questions and problems is given. In particular, methods for the simulation of the typical cell for other types of random tessellation models like the aggregated Poisson–Voronoi tessellation are discussed. Possible extensions of the described methods for cost analysis of two–level hierarchical models, for example, with respect to the determination of the distribution of the used characteristics, are explained. A short outlook is given with regard to dynamic modelling and applications for dynamics as well as for cost analysis in cell biology.

This thesis is wrapped up by an appendix that introduces some useful basic mathematical definitions. In the first two parts some notions from set theory and topology like sets, metric spaces and denseness are given. In the last two parts basic concepts from measure theory and probability calculus like measures, distributions and the central limit theorem are explained.

1.3   The GeoStoch Library

The software developed in the course of this thesis is embedded in the GeoStoch library which is a joint project between the Institute of Applied Information Processing and the Institute of Stochastics at Ulm University ([62], [65]). This JAVA–based software library comprises methods from stochastic geometry, spatial statistics and image analysis.

The basic idea behind this software project is to offer a core of general methods that are useful for different kinds of applications thereby ensuring a high degree of reusability. Additionally, the library is constantly extended in the course of the various research projects, for example in the cooperation with France Télécom R&D and in the projects concerning applications in cell biology. Further information about the GeoStoch software library and some of the projects that are applying the software can be found on its internet domain *http://www.geostoch.de.*

Chapter 2

Random Tessellations and Basics from Stochastic Geometry

The aim of this chapter is to introduce the notion of a random tessellation in $\mathbb{R}^2$ and to briefly discuss some particular tessellation models as well as to study their properties. Note that in the following we will concentrate on $\mathbb{R}^2$, but most of the definitions given here can be canonically extended to $\mathbb{R}^d$.

In order to define random tessellations in $\mathbb{R}^2$ some basic concepts of stochastic geometry have to be introduced first. Therefore, in the first parts of this chapter (Sections 2.1 and 2.2) we define random closed sets and random point processes. Topics like Palm distributions and Neveu's exchange theorem that will become important in Chapter 4 are also discussed there. Afterwards, in Section 2.3 a particuar model for a random closed set, the Boolean model, is introduced in order to define so–called modulated Poisson point processes based on these Boolean models. Finally, in Section 2.4 random tessellations are defined and numerous examples, like Poisson–Voronoi tessellations, modulated tessellations and iterated tessellations are provided.

For a detailed introduction to the development of stochastic geometry as well as for a positioning among closely related fields like integral and convex geometry, spatial statistics, and stereology we refer to [94]. With respect to even more basic concepts from set theory, topology, measure theory and probability calculus have a look at the appendix and the references therein. More informations on the topics discussed in this chapter can be found in the existing literature. In particular, we refer to [70] and [69] for a study of random closed sets and more specifically of the Boolean model. Random point processes are discussed in [22], [46], [47], [61], and [108], whereas a profound discussion of stochastic geometry including models for more general settings can be found in [2], [12], [37], [44], [68], [91], [94], [100], and [101]. Specific information about random tessellations and their properties are, for example, given in [68], [72], [71], [78], [94], and [100]. Finally we mention [55] which, without neglecting the work and contributions of many mathematicians in former times, can be regarded as the starting point of stochastic geometry as a mathematical discipline of its own right.

2.1   Random Closed Sets

One of the key elements of stochastic geometry are random closed sets. Examples of such random closed sets are given by random point patterns and by balls with random radius and random center points, respectively. We consider the two–dimensional Euclidean space $\mathbb{R}^2$ equipped with the Borel–σ–algebra $\mathcal{B}(\mathbb{R}^2)$, where $\mathcal{B}_0(\mathbb{R}^2)$ denotes the system of all bounded Borel sets in $\mathbb{R}^2$. The two–dimensional Lebesgue measure is denoted by $\nu_2(B)$ for any $B \in \mathcal{B}(\mathbb{R}^2)$. Furthermore, let $\mathcal{F}$, $\mathcal{C}$ and $\mathcal{K}$ be the family of all closed sets, the family of all compact sets, and the family of all convex bodies, respectively. With respect to $\mathcal{F}$ we define the Borel–σ–algebra $\mathcal{B}(\mathcal{F})$ as the smallest σ–algebra of $\mathcal{F}$ containing all sets $\{F \in \mathcal{F}, F \cap C \neq \emptyset\}$ for arbitrary $C \in \mathcal{C}$. By $(\Omega, \mathcal{A}, \mathbb{P})$ we denote some probability space.

A $(\mathcal{A}, \mathcal{B}(\mathcal{F}))$–measurable mapping $\Xi : \Omega \to \mathcal{F}$ which maps the probability space $(\Omega, \mathcal{A}, \mathbb{P})$ into the space $(\mathcal{F}, \mathcal{B}(\mathcal{F}))$ is called a *random closed set* in $\mathbb{R}^2$. Additionally, if $\mathbb{P}(\Xi \in \mathcal{C}) = 1$ the set Ξ is called a *random compact set*. By $P_\Xi : \mathcal{B}(\mathcal{F}) \to [0,1]$ we denote the *distribution* of the random closed set Ξ, where P_Ξ is given by

$$P_\Xi(B) = \mathbb{P}(\Xi \in B) = \mathbb{P}(\{\omega \in \Omega : \Xi(\omega) \in B\})$$

for any $B \in \mathcal{B}(\mathcal{F})$. A random closed set Ξ is said to be

- *stationary* if its distribution P_Ξ is invariant under translation, i.e., $P_{\Xi+x} = P_\Xi$ for any $x \in \mathbb{R}^2$, where $\Xi + x$ is the random closed set Ξ shifted by the vector x,

- *isotropic* if P_Ξ is invariant under rotation, i.e., $P_{\vartheta(\Xi)} = P_\Xi$ for any rotation $\vartheta : \mathbb{R}^2 \to \mathbb{R}^2$ around the origin o, and

- *motion–invariant* if Ξ is both stationary and isotropic.

In Section 4.3 the coverage probability p_Ξ of a random closed set Ξ will play an important role. For a stationary random closed set Ξ the coverage probability p_Ξ is defined as

$$p_\Xi = \mathbb{P}(o \in \Xi). \tag{2.1}$$

Note that $p_\Xi = \mathbb{E}\left[\nu_2\left(\Xi \cap [0,1]^2\right)\right]$. In order to introduce ergodicity of a random closed set Ξ we first have to define the notion of an averaging sequence. Hence a sequence of bounded Borel sets $B_1, B_2, \ldots \in \mathcal{B}_0(\mathbb{R}^2)$ is called an *averaging sequence* if the following three conditions hold

- B_n is convex for all $n \geq 1$,

- $B_n \subset B_{n+1}$ for all $n \geq 1$,

- $\lim_{n \to \infty} \rho(B_n) = \infty$, where $\rho(B) = \sup\{r \geq 0 : b(x, r) \subseteq B, x \in B\}$.

A stationary random closed set $\Xi \subset \mathbb{R}^2$ with distribution P_Ξ is called *ergodic* if for every averaging sequence $B_1, B_2, \ldots \in \mathcal{B}_0(\mathbb{R}^2)$ and for all $A, A' \in \mathcal{B}(\mathcal{F})$

$$\lim_{n \to \infty} \frac{1}{\nu_2(B_n)} \int_{B_n} (P_\Xi(A_x \cap A') - P_\Xi(A)P_\Xi(A'))dx = 0,$$

where $A_x = \{y + x, \ y \in A\}$ is the set A shifted by the vector $x \in \mathbb{R}^2$.

2.2 Random Point Processes

Random point processes in $\mathbb{R}^2$ are a very common model of stochastic geometry for points randomly scattered in the plane. In this thesis we will concentrate on simple point processes in the plane, i.e., point processes where at each location at most one point can be located. Such point processes are used, for example, to model point patterns that arise in various fields of economy and science like biology, forestry, medicine, and telecommunication ([10], [25], [27], [57], [58], [59]).

2.2.1 Definition of Random Point Processes

A point $x \in \mathbb{R}^2$ can be described by a *point measure* δ_x that is defined with respect to $A \in \mathcal{B}(\mathbb{R}^2)$ by

$$\delta_x(A) = \begin{cases} 1 & \text{if } x \in A, \\ 0 & \text{if } x \notin A. \end{cases}$$

Hence δ_x is a probability measure on the measurable space $(\mathbb{R}^2, \mathcal{B}(\mathbb{R}^2))$. A finite (or countably infinite) sum

$$\varphi = \sum_{i=1}^{k} \delta_{x_i},$$

where $k \in \mathbb{N}_0 \cup \{\infty\}$, of such point measures is also a measure on $(\mathbb{R}^2, \mathcal{B}(\mathbb{R}^2))$. This measure φ is a counting measure with non–negative integer values where the value '∞'

is possible as well. The points $x \in \mathbb{R}^2$ with $\varphi(\{x\}) > 0$ are called *atoms* of the counting measure φ. The set $S_\varphi = \{x \in \mathbb{R}^2 : \varphi(\{x\}) > 0\}$ of the atoms of φ is called the *support* of φ. Later on we will concentrate on *simple* counting measures, where $\varphi(\{x\}) \leq 1$ for all $x \in \mathbb{R}^2$. Furthermore, we will only regard *locally finite* counting measures which means that to every bounded Borel set $B \in \mathcal{B}_0(\mathbb{R}^2)$ a non–negative integer value $\varphi(B)$ is assigned, where

$$\varphi(B) = \sum_{x \in S_\varphi} \varphi(\{x\})\delta_x(B).$$

Let N denote the set of all non–negative locally finite counting measures $\varphi : \mathcal{B}(\mathbb{R}^2) \rightarrow \mathbb{N}_0 \cup \{\infty\}$ and let $\mathcal{N}$ denote the smallest σ–algebra of subsets of N that contains all sets of the form $\{\varphi \in N : \varphi(B) = k\}$, where $k \in \mathbb{N}_0$ and where $B \in \mathcal{B}_0(\mathbb{R}^2)$ is an arbitrary bounded Borel set in $\mathbb{R}^2$.

A *random point process* X in $\mathbb{R}^2$ is a $(\mathcal{A}, \mathcal{N})$–measurable mapping $X : \Omega \rightarrow N$ that is defined on the probability space $(\Omega, \mathcal{A}, \mathbb{P})$ and that has values in the measurable space $(N, \mathcal{N})$. The *distribution* P_X of X is given by

$$P_X(A) = \mathbb{P}(X \in A) = \mathbb{P}(\{\omega \in \Omega : X(\omega) \in A\}),$$

for any $A \in \mathcal{N}$.

The number of points of a random point process X in a set $B \in \mathcal{B}(\mathbb{R}^2)$ is denoted by $X(B)$. Note that we will sometimes omit the notion random point process and instead only write point process without changing the meaning.

Often it is useful to regard an alternative representation of a random point process X which is given by $X = \{X_n\}_{n \geq 1}$. In this case, the point process X is considered as a sequence of random points $X_1, X_2, \ldots$ instead of a measurable mapping into the measurable space $(N, \mathcal{N})$. From this point of view, we can regard (simple) random point processes as a special case of random closed sets introduced in Section 2.1.

2.2.2   Properties of Random Point Processes

A random point process X in $\mathbb{R}^2$ is called

1. *stationary* if its distribution P_X is invariant under translation, i.e., $P_{T_x X} = P_X$ for any $x \in \mathbb{R}^2$, where $T_x X(B) = X(B_{-x})$ for any $B \in \mathcal{B}(\mathbb{R}^2)$, with B_{-x} denoting the set B translated by the vector $-x \in \mathbb{R}^2$,

2. *isotropic* if its distribution is invariant under rotation, i. e., $P_{\vartheta(X)} = P_X$ for any rotation ϑ around the origin o, and

3. *motion–invariant* if X is both stationary and isotropic.

A basic characteristic for any random point process X in $\mathbb{R}^2$ is its *intensity measure* $\Lambda_X : \mathcal{B}(\mathbb{R}^2) \to [0, \infty]$ that is defined by

$$\Lambda_X(B) = \mathbb{E}X(B) = \int_N \varphi(B) P_X(d\varphi),$$

for any $B \in \mathcal{B}(\mathbb{R}^2)$. In the following we will only regard random point processes with a locally finite intensity measure that is not equal to the zero measure, i.e., $\Lambda_X(B) < \infty$ for any bounded $B \in \mathcal{B}_0(\mathbb{R}^2)$ and $\Lambda_X(\mathbb{R}^2) > 0$.

If X is stationary we have that

$$\Lambda_X(B) = \mathbb{E}X(B) = \mathbb{E}T_x X(B) = \mathbb{E}X(B_{-x}) = \Lambda_X(B_{-x})$$

for arbitrary $x \in \mathbb{R}^2$, $B \in \mathcal{B}(\mathbb{R}^2)$, where B_{-x} denotes the set B translated by the vector $-x \in \mathbb{R}^2$. Now, due to the fact that every stationary Radon measure on $(\mathbb{R}^2, \mathcal{B}(\mathbb{R}^2))$, i.e., every Borel measure which is finite on compact sets, is a multiple of the Lebesgue measure (cmp. Lemma A.1) the intensity measure Λ_X can be expressed as

$$\Lambda_X(B) = \lambda_X \nu_2(B),$$

for any $B \in \mathcal{B}(\mathbb{R}^2)$. The constant λ_X is called the *intensity* of the point process X. It can be interpreted as the mean number of points of X per unit area, i.e., $\lambda_X = \mathbb{E}X([0,1)^2)$. Due to the fact that Λ_X is locally finite and not equal to the zero measure we have that $\lambda_X \in (0, \infty)$.

A property of random point processes that will often be used is ergodicity. A random point process X in $\mathbb{R}^2$ with distribution P_X is said to be *ergodic* if for any averaging sequence (cmp. Section 2.1) $B_1, B_2, \ldots \in \mathcal{B}_0(\mathbb{R}^2)$ and for all $A, A' \in \mathcal{N}$

$$\lim_{n \to \infty} \frac{1}{\nu_2(B_n)} \int_{B_n} (P_X(T_x A \cap A') - P_X(A)P_X(A'))dx = 0. \qquad (2.2)$$

In Sections 2.2.4 and 2.2.5 we will see examples for ergodic as well as for non–ergodic random point processes. The property of ergodicity is useful in order to ensure that statistical averages can be expressed by limits of spatial averages and vice versa. In other words, if the point process is ergodic there is in a sense no difference between looking at means with respect to different realisations or with respect to a large sampling window (cmp. [100], p. 198).

Another important property that is related to stationarity is the following. We call a stationary random point process X *mixing* if for all $A, A' \in \mathcal{N}$

$$P_X(T_x A \cap A') - P_X(A)P_X(A') \to 0 \qquad (2.3)$$

for $|x| \to \infty$, where $|\cdot|$ denotes the Euclidean norm. Furthermore, the following theorem holds that describes a relationship between the ergodicity and the mixing property of a random point process X.

Theorem 2.1 *Let X be a random point process in $\mathbb{R}^2$. If X is stationary and mixing then X is ergodic.*

Proof   Consider a stationary and mixing point process X in $\mathbb{R}^2$. Due to the mixing property of X we have that for all $\varepsilon > 0$ there is an $R > 0$ such that for all $x \in \mathbb{R}^2$ with $|x| > R$ it holds that $|P_X(T_xA \cap A') - P_X(A)P_X(A')| < \varepsilon$. By looking at the definition of ergodicity we obtain that

$$\lim_{n \to \infty} \frac{1}{\nu_2(B_n)} \int_{B_n} (P_X(T_xA \cap A') - P_X(A)P_X(A'))dx$$

$$= \lim_{n \to \infty} \frac{1}{\nu_2(B_n)} \int_{B_n \cap \{x:|x| \leq R\}} (P_X(T_xA \cap A') - P_X(A)P_X(A'))dx$$

$$+ \lim_{n \to \infty} \frac{1}{\nu_2(B_n)} \int_{B_n \setminus \{x:|x| \leq R\}} (P_X(T_xA \cap A') - P_X(A)P_X(A'))dx$$

Due to the fact that $\lim_{n \to \infty} \int_{B_n \cap \{x:|x| \leq R\}} (P_X(T_xA \cap A') - P_X(A)P_X(A'))dx < \infty$ and that $\nu_2(B_n) \to \infty$ for $n \to \infty$, the first summand converges towards 0 as $n \to \infty$. The second summand also converges towards 0 as $n \to \infty$, because it is only integrated with respect to $x \in B_n$ with $|x| > R$. Due to the assumption that X is mixing, we therefore get that the second summand becomes arbitrarily small. In summary, we obtain that X is ergodic. $\qquad\qquad\square$

In order to check for ergodicity and for the mixing property the notion of a semiring proves to be a useful tool. A *semiring* in $(N, \mathcal{N})$ is a non–empty family $\mathcal{J} \subset \mathcal{N}$ of subsets of N such that

- if $E_1, ..., E_n \in \mathcal{J}$ then it follows that $\cap_{k=1}^n E_k \in \mathcal{J}$,

- if $E, F \in \mathcal{J}$ and $E \subset F$ then there exists a finite sequence $\{C_1, ..., C_n\}$ of sets in $\mathcal{J}$ such that $E = C_0 \subset C_1 \subset ... \subset C_n = F$ and $D_i = C_i \setminus C_{i-1} \in \mathcal{J}$ for all $i = 1, ..., n$.

Note that the family of all subsets of N that have the form

$$\{\varphi \in N : \varphi(A_i) \in B_i, \ i = 1, ..., m\} \tag{2.4}$$

for $A_i \in \mathcal{B}_0(\mathbb{R}^2)$, $B_i \in \mathcal{B}(\mathbb{R})$ and $m \in \mathbb{N}$, is a semiring in $(N, \mathcal{N})$ which is generating the σ–algebra $\mathcal{N}$.

With respect to ergodicity and mixing properties we can use the semiring described by (2.4) in the following two lemmas.

Lemma 2.1 *A stationary point process is ergodic if and only if* (2.2) *holds for any* $A, A' \in \mathcal{J}$, *where* $\mathcal{J}$ *is a semiring which generates* $\mathcal{N}$.

Lemma 2.2 *A stationary point process is mixing if and only if* (2.3) *holds for any* $A, A' \in \mathcal{J}$, *where* $\mathcal{J}$ *is a semiring which generates* $\mathcal{N}$.

Proofs for Lemmas 2.1 and 2.2 are given in [22] on p. 342.

2.2.3 Palm Distributions and Campbell Measures

With regard to the analysis of random point processes it is often convenient to express the distribution of a point process X in $\mathbb{R}^2$ conditional to the event that a point of X is located at a specific location $x \in \mathbb{R}^2$. For example, an interesting probability is of the form $\mathbb{P}(X(b(x, r)) = 1 \mid x \in X)$. In other words we want to look at the probability that in the disc around a location x there is no other point of X under the condition that x belongs to the point process X. Obviously it is not possible to express such a probability as a conditional probability in the usual sense since in most cases $\mathbb{P}(x \in X) = 0$. Hence, there is a need to define an alternative approach in order to attack this problem. Such an approach is given by the *Palm distribution* which will be introduced in the following.

Let X be a stationary random point process in $\mathbb{R}^2$ with finite intensity $\lambda_X > 0$ and $B \in \mathcal{B}(\mathbb{R}^2)$ with $0 < \nu_2(B) < \infty$. Then we call the set function $P_X^* : \mathcal{N} \to [0, 1]$ defined by

$$P_X^*(A) = \frac{1}{\lambda_X \nu_2(B)} \mathbb{E} \sum_{x \in S_X \cap B} X(\{x\}) \mathbb{1}_A(T_{-x}X), \ A \in \mathcal{N} \qquad (2.5)$$

the *Palm distribution* (with respect to the origin o) of X.

If X is an ergodic point process then its Palm distribution $P_X^*(A)$ may be interpreted as the probability that a typical point $x \in X$ has the property that $T_{-x}X$ belongs to $A \in \mathcal{N}$. It can be shown that P_X^* has indeed the properties of a distribution function and that, due to the stationarity of X, the Palm distribution is independent from the choice of $B \in \mathcal{B}(\mathbb{R}^2)$. It is also possible to define a Palm distribution P_X^* with respect to a general location $x \in \mathbb{R}^2$ which is not necessarily the origin.

Note that another introduction of the Palm distribution P_X^* of X can be given as the Radon density of the Campbell measure with respect to the intensity measure at the location o.

The *Campbell measure* $C_X : \mathcal{N} \times \mathcal{B}(\mathbb{R}^2) \to [0, \infty]$ is defined by

$$C_X(A \times B) = \int_N \varphi(B) \mathbb{1}_A(\varphi) P_X(d\varphi) = \int_N \sum_{x \in S_\varphi} \varphi(\{x\}) \mathbb{1}_B(x) \mathbb{1}_A(\varphi) P_X(d\varphi) \qquad (2.6)$$

Note that the Campbell measure can be regarded as a refinement of the intensity measure Λ_X since $\Lambda_X(B) = C_X(N \times B)$ for all $B \in \mathcal{B}(\mathbb{R}^2)$. Due to the fact that the intensity measure is locally finite we obtain the following decomposition of C_X.

Theorem 2.2 *For almost all (with respect to the intensity measure Λ_X) $x \in \mathbb{R}^2$ there exists a uniquely determined distribution $P_X^{(x)}$ on $\mathcal{N}$ such that*

$$C_X(A \times B) = \int_B P_X^{(x)}(A) \Lambda_X(dx) \qquad (2.7)$$

for all $A \in \mathcal{N}$ and for all $B \in \mathcal{B}(\mathbb{R}^2)$.

For a proof of Theorem 2.2 have a look at [46], pp. 60 and 308. Note that in this way a general Palm distribution $P_X^{(x)}$ is defined that is not limited to stationary point processes. For a stationary point process it holds that $P_X^{(x)}(A) = P_X^*(A_x)$ for any $A \in \mathcal{N}$.

The definition of the Campbell measure provided by (2.6) can be seen as a special case of a more general relationship between the measures P_X^* and C_X given by the *refined Campbell theorem*.

Theorem 2.3 *(Refined Campbell theorem) Let X be a stationary point process in $\mathbb{R}^2$ with a finite intensity $\lambda_X > 0$ and $f : N \times \mathbb{R}^2 \to [0, \infty)$ an arbitrary measurable function. Then it holds that*

$$\mathbb{E} \sum_{x \in S_X} X(\{x\}) f(X, x) = \lambda_X \int_{\mathbb{R}^2} \mathbb{E}_{P_X^*} f(T_{-x}X, x) dx, \qquad (2.8)$$

where $\mathbb{E}_{P_X^*}$ is the expectation with respect to P_X^*.

For a proof of Theorem 2.3 for instance have a look at Sections 4.1 and 7.1 of [100] and Chapter 4 of [94]. An immediate consequence of the refined Campbell theorem is given by the Campbell theorem that is essential for working with point processes.

Corollary 2.3 *(Campbell theorem) Let X be a stationary point process in $\mathbb{R}^2$ with a finite intensity $\lambda_X > 0$ and $f : \mathbb{R}^2 \to [0, \infty)$ an arbitrary measurable function. Then it holds that*

$$\mathbb{E} \sum_{x \in S_X} X(\{x\}) f(x) = \lambda_X \int_{\mathbb{R}^2} f(x) dx. \tag{2.9}$$

2.2.4 Poisson Point Processes

The Poisson point process is the most prominent example of a point process model. It reflects the state of complete spatial randomness, where all points are scattered conditionally uniformly and independently of each other on the plane. Therefore, it is often used as a reference model or as a basis for the construction of more sophisticated models like the Matern–cluster model or the Boolean model that will be introduced in Sections 2.2.5 and 2.3, respectively. In this section the main characteristics of a Poisson point process are described.

Let X denote a point process in $\mathbb{R}^2$. We call X a *Poisson point process* in $\mathbb{R}^2$ with intensity measure Λ_X if the following two properties are fulfilled

1. Let $B \in \mathcal{B}_0(\mathbb{R}^2)$ be any bounded Borel set. The number of points $X(B)$ of X occurring in B is Poisson distributed with mean $\Lambda_X(B)$, i.e.,

$$\mathbb{P}(X(B) = n) = \frac{\Lambda_X(B)^n}{n!} \exp(-\Lambda_X(B)), \ n \in \mathbb{N}_0.$$

2. Let $B_1, ... B_k \in \mathcal{B}(\mathbb{R}^2)$, $k \in \mathbb{N}$, denote k pairwise disjoint Borel sets. The random variables $X(B_1), ..., X(B_k)$ are mutually independent for all $k \in \mathbb{N}$.

In the case of a stationary Poisson point process X (as for any stationary point process) we have that $\Lambda_X(B) = \lambda_X \nu_2(B)$ for an intensity $\lambda_X \in (0, \infty)$. Often, even in instationary cases, it is possible to express Λ_X with respect to the Lebesgue measure via a density. This means that there exists a Borel–measurable mapping $\lambda_X : \mathbb{R}^2 \to [0, \infty)$ such that

$$\Lambda_X(B) = \int_B \lambda_X(x) dx, \text{ for all } B \in \mathcal{B}(\mathbb{R}^2). \tag{2.10}$$

Here, $\lambda_X(x)$ can be considered as the local intensity of X at a location $x \in \mathbb{R}^2$. In Figure 2.1 realisations of an example for a stationary as well as an example for an instationary Poisson point process are displayed. The *void probabilities* of stationary Poisson point processes are given by

$$\mathbb{P}(X(B) = 0) = \exp(-\lambda_X \nu_2(B))$$

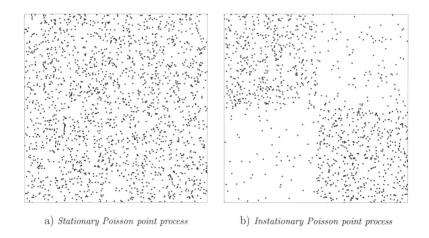

a) *Stationary Poisson point process*   b) *Instationary Poisson point process*

Figure 2.1: Realisations of Poisson point processes

for any $B \in \mathcal{B}(\mathbb{R}^2)$ which follows directly from the first property of the Poisson point process mentioned above by putting $n = 0$.

A stationary Poisson point process as defined above is always isotropic and therefore motion–invariant. These properties may be verified by noting that the distributions remain the same whether one uses a stationary Poisson point process X or its translation $T_x X$ or its rotation $\vartheta(X)$ around the origin.

Lemma 2.4 *A stationary Poisson point process X is mixing and therefore ergodic.*

Proof   Let X be a stationary Poisson point process. Furthermore let $A, A' \in \mathcal{J}$, where $\mathcal{J}$ is the semiring given in (2.4). For $x \in \mathbb{R}^2$ with $|x|$ sufficiently large we obtain that

$$P_X(T_x A \cap A') = P_X(T_x A) P_X(A') = P_X(A) P_X(A').$$

Therefore, the stationary Poisson point process X is mixing and due to Theorem 2.1 also ergodic.   □

2.2.5   Other Examples of Random Point Processes

Apart from the Poisson point process defined in Section 2.2.4, there are various other models for random point processes. Some of them are introduced here.

The most simpliest point process consists only of a single (deterministic) point x. This (degenerate) point process will be denoted by δ_x in the following. It will later be useful, for example, in the representation of Slivnyak's theorem (Section 3.1.2).

Based on the Poisson point process some related random point processes X can be constructed. A first possibility is to take the intensity measure Λ_X itself random. For this purpose we consider N', the set of all locally finite measures $\eta : \mathcal{B}(\mathbb{R}^2) \rightarrow [0, \infty]$, together with its corresponding σ–algebra $\mathcal{N}'$ which is the smallest σ–algebra such that $\eta \rightarrow \eta(B)$ is a $(\mathcal{N}', \mathcal{B}(\mathbb{R}))$–measurable mapping for any $B \in \mathcal{B}(\mathbb{R}^2)$. We call a measurable mapping $\Lambda : \Omega \rightarrow N'$ a *random measure* if it is a mapping from a probability space $(\Omega, \mathcal{A}, \mathbb{P})$ into the measurable space $(N', \mathcal{N}')$. This generalization leads to *doubly stochastic Poisson processes* or *Cox point processes*. More formally, consider an arbitrary locally finite random measure Λ_X. We call X a Cox point process with random intensity measure Λ if

$$\mathbb{P}(\bigcap_{i=1}^{n} \{X(B_i) = k_i\}) = \mathbb{E}\left(\prod_{i=1}^{n} \frac{\Lambda^{k_i}(B_i)}{k_i!} \exp(-\Lambda^{k_i}(B_i)) \right) \qquad (2.11)$$

for any $n \geq 1$, $k_1, ..., k_n \geq 0$ and for pairwise disjoint $B_1, ..., B_n \in \mathcal{B}_0(\mathbb{R}^2)$. Note that the distribution $P_X : \mathcal{N} \otimes \mathcal{N}' \rightarrow [0, 1]$ of X is induced by the distribution $P_\Lambda : \mathcal{N}' \rightarrow [0, 1]$, where

$$\mathbb{P}(\bigcap_{i=1}^{n} \{X(B_i) = k_i\}) = \int_{N'} \prod_{i=1}^{n} \frac{\eta^{k_i}(B_i)}{k_i!} \exp(-\eta^{k_i}(B_i)) P_{\Lambda_X}(d\eta). \qquad (2.12)$$

Hence we are able to describe the distribution of the Cox point process X as a mixture of the distributions of (not necessarily stationary) Poisson point processes. Furthermore, we can think of a Cox point process as a two–step random mechanism. In a first step a realisation η of the random measure Λ is determined according to a distribution P_Λ. Afterwards, in a second step, a Poisson point process is generated according to the intensity measure η. Some interesting examples of Cox point processes will be discussed in the course of this thesis, for example in Sections 2.3.2 and 2.4.6.

From the definition of a Cox point process given in (2.11) we can directly deduce the following lemma which will be useful in Section 2.3.2.

Lemma 2.5 *Let X be a Cox point process with stationary random intensity measure Λ_X. Then X is a stationary point process.*

A prominent example of a Cox point process that is also an example of a non–ergodic point process is the *mixed Poisson point process*. It is defined as a stationary Poisson point process X, where the intensity λ_X is randomly chosen. Therefore, the random driving measure can be described as

$$\Lambda_X(B) = Y\nu_2(B)$$

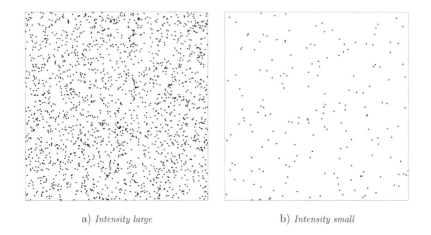

a) *Intensity large*   b) *Intensity small*

Figure 2.2: Two realisations of a mixed Poisson point process

for any $B \in \mathcal{B}(\mathbb{R}^2)$, where Y is some non–negative random variable on $\mathbb{R}$. By checking the definition for ergodicity of a random point process given in Section 2.2.2 it can be seen that a mixed Poisson point process is non–ergodic (see also [100] on p. 154). A consequence of this fact is that spatial averages over a single realisation are not comparable to arithmetical averages over a certain number, say n, realisations. Two realisations of a specific mixed Poisson point process, where Y can only have two different values are shown in Figure 2.2. Note that a realisation of a mixed Poisson point process always looks like a realisation of a stationary (and therefore ergodic) Poisson point process with the corresponding (deterministic) intensity.

Other important examples for random point processes whose construction principles are based on the Poisson point process are *Neyman–Scott processes* and *Matern hardcore processes*. The general model of a *Neyman–Scott point process* is provided by a stationary Poisson point process whose points act as the parent points of the Neyman–Scott process. Around each parent point, the points of the Neyman–Scott process are scattered independently and with identical distribution. A special case of the Neyman–Scott process is the *Matern–cluster process*, where the points are scattered around a parent point according to another Poisson point process inside a disc with a fixed radius R. A realisation of a Matern–cluster process can be seen in Figure 2.3a. Such a model is often useful in order to model point processes that show a clustering effect, i.e., for distances below a certain value there is an attraction between point pairs of

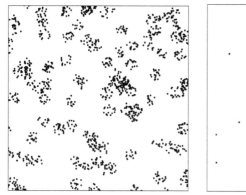

a) *Realisation of a Matern–cluster process*   b) *Realisation of a Matern hardcore process*

Figure 2.3: Matern–cluster process and Poisson hardcore process

such a distance, whereas for larger distances the points behave independently (cmp. e.g. [25], [27]).

If, instead of a clustering effect, a hardcore effect is observed the *Matern hardcore process* might prove to be a useful model. Here, all the points belonging to the point process have at least a certain minimal distance (the hardcore distance) to each other. Apart from this property the points are scattered conditionally uniformly (cmp. e..g [100], pp. 162–166). A realisation of a Matern hardcore process can be seen in Figure 2.3b.

2.2.6 Marked Point Processes

Marked point processes can be considered as a generalisation of random point processes. We want to regard marked point processes as random counting measures on $\mathbb{R}^2$ that are equipped with a mark from the *mark space* $\mathbb{M}$. In the following $\mathbb{M}$ denotes an arbitrary Polish space and $\mathcal{M}$ the σ–algebra of the Borel sets of $\mathbb{M}$.

For the definition of a random marked point process, regard the set $N_\mathbb{M} = N(\mathbb{R}^2 \times \mathbb{M})$ of all measures $\psi : \mathcal{B}(\mathbb{R}^2) \otimes \mathcal{M} \to \mathbb{N}_0 \cup \{\infty\}$ which are locally finite and simple in the first component. Therefore, each measure $\psi \in N_\mathbb{M}$ can be written in the form

$$\psi(B \times L) = \sum_{(x,m) \in S_\psi} \delta_{[x,m]}(B \times L), \qquad (2.13)$$

where $B \in \mathcal{B}(\mathbb{R}^2)$, $L \in \mathcal{M}$ and where S_ψ denotes the *support* of ψ. So, $\psi(B \times L)$ denotes the number of points x that are located in B and have a mark belonging to L. Let $\mathcal{N}_M = \mathcal{N}(\mathbb{R}^2 \times \mathbb{M})$ be the smallest σ–algebra containing all subsets of N_M that are of the form $\{\psi \in N_M : \psi(B \times L) = k\}$, with $B \in \mathcal{B}_0(\mathbb{R}^2)$, $L \in \mathcal{M}$, and $k \in \mathbb{N}_0$. A *random marked point process* X_D in $\mathbb{R}^2$ with mark space $(\mathbb{M}, \mathcal{M})$ is given by the mapping $X_D : \Omega \to N_M$, that maps the probability space $(\Omega, \mathcal{A}, \mathbb{P})$ into the measurable space $(N_M, \mathcal{N}_M)$.

The *distribution* P_{X_D} of a marked point process X_D is given by

$$P_{X_D}(A) = \mathbb{P}(X_D \in A) = \mathbb{P}(\{\omega \in \Omega : X_D(\omega) \in A\}) \qquad (2.14)$$

for any $A \in \mathcal{N}_M$. Analogously to random point processes without marks, we can write the marked point process X_D as $X_D = \{(X_n, D_n)\}_{n \geq 1}$ with the measurable mappings $X_n : \Omega \to \mathbb{R}^2$ and $D_n : \Omega \to \mathbb{M}$. This means that instead of representing X_D as a measurable mapping into the space $(N_M, \mathcal{N}_M)$, we can regard it as a tupel of two sequences $\{X_n\}_{n \geq 1}$ and $\{D_n\}_{n \geq 1}$, where $\{X_n\}_{n \geq 1}$ is a sequence of random (unmarked) points and $\{D_n\}_{n \geq 1}$ is a sequence of random marks. The marked point process X_D is called *independently marked* if the sequences $\{X_n\}_{n \geq 1}$ and $\{D_n\}_{n \geq 1}$ are independent and additionally, if $\{D_n\}_{n \geq 1}$ consists of independent and identically distributed random variables.

A marked point process $X_D = \{[X_n, D_n]\}_{n \geq 1}$ is called

- *stationary* if $P_{X_D} = P_{T_x X_D}$, where $T_x X_D = \{(X_n + x, D_n)\}_{n \geq 1}$ for any $x \in \mathbb{R}^2$.

- *isotropic* if $P_{X_D} = P_{\vartheta(X_D)}$ for all rotations ϑ around the origin o, where $\vartheta(X_D) = \{\vartheta(X_n), D_n\}_{n \geq 1}$.

- *motion–invariant* if X_D is both stationary and isotropic.

The mapping $\Lambda_{X_D} : \mathcal{B}(\mathbb{R}^2) \otimes \mathcal{M} \to [0, \infty]$ given by

$$\Lambda_{X_D}(B \times L) = \mathbb{E}\left(X_D(B \times L)\right), \ B \in \mathcal{B}(\mathbb{R}^2), \ L \in \mathcal{M} \qquad (2.15)$$

is called the *intensity measure* of X_D, where $X_D(B \times L)$ denotes the number of (marked) points of X_D located in $B \times L$. Note that similarly to unmarked stationary point processes we can define an intensity λ_{X_D} for stationary marked point processes by taking $L = \mathbb{M}$ in (2.15).

A stationary marked point process X_D with distribution P_{X_D} is said to be *ergodic* if for any averaging sequence $B_1, B_2, ... \in \mathcal{B}_0(\mathbb{R}^2)$ and for all $A, A' \in \mathcal{N}_M$

$$\lim_{n \to \infty} \frac{1}{\nu_2(B_n)} \int_{B_n} (P_{X_D}(T_x A \cap A') - P_{X_D}(A)P_{X_D}(A'))dx = 0. \qquad (2.16)$$

We call a stationary marked point process X_D *mixing* if for all $A, A' \in \mathcal{N}_M$

$$P_{X_D}(T_x A \cap A') - P_{X_D}(A) P_{X_D}(A') \to 0, \text{ for } |x| \to \infty. \qquad (2.17)$$

2.2.7 Neveu's Exchange Formula for Palm Distributions

In this subsection Neveu's exchange formula adapted to (marked) point process in $\mathbb{R}^2$ is presented. This formula allows to express the relationship of expectations for functionals of two stationary marked point processes X_D and $\widetilde{X}_{\widetilde{D}}$ with respect to their Palm distributions $P^*_{X_D}$ and $P^*_{\widetilde{X}_{\widetilde{D}}}$, respectively.

In order to capture the randomness of such a system of several random processes, we consider a *flow* $\{\theta_x : x \in \mathbb{R}^2\}$ on the space Ω, i.e., a family of bijective shift operators $\theta_x : \Omega \to \Omega$ such that $\theta_x \circ \theta_y = \theta_{x+y}$, where $\circ$ denotes the concatenation operator. We additionally assume that the mapping $f : \mathbb{R}^2 \times \Omega \to \Omega$ with $f(x, \omega) = \theta_x \omega$ is measurable. For $x \in \mathbb{R}^2$ let θ_x be compatible with the shift operator T_x as defined above, which means that

$$X_D(\theta_x \omega, B \times L) = T_x X_D(\omega, B \times L) = X_D(\omega, B_{-x} \times L), \qquad (2.18)$$

for a given marked point process X_D and all $B \in \mathcal{B}(\mathbb{R}^2)$, $L \in \mathcal{M}$. Note that we can obtain the stationarity of X_D by assuming that

$$P_{X_D}(\theta_x A) = P_{X_D}(\theta_x^{-1} A) = P_{X_D}(A), \qquad (2.19)$$

for all $A \in \mathcal{A}$ and $x \in \mathbb{R}^2$, where $\theta_x A = \{\theta_x \omega : \omega \in A\}$.

Also, using the definition of the operator θ_x, we introduce the *Palm distribution* $P^*_{X_D}$ for a stationary marked point process X_D as a probability measure on the product σ–algebra $\mathcal{A} \otimes \mathcal{M}$ given by

$$P^*_{X_D}(A \times L) = \frac{1}{\lambda \nu_2(B)} \int_\Omega \int_{\mathbb{R}^2 \times L} \mathbb{I}_B(x)\, \mathbb{I}_A(\theta_x \omega)\, X(\omega, d(x, g)) \mathbb{P}(d\omega) \qquad (2.20)$$

for any $B \in \mathcal{B}(\mathbb{R}^2)$ with $0 < \nu_2(B) < \infty$. In (2.20) the Palm distribution is defined as a probability measure with respect to a set $A \times L$, where $A \in \mathcal{A}$. On first sight, this seems to be contrary to the definition of the Palm distribution given in 2.5 with respect to a set $A \in \mathcal{N}$. But note that we are able to transfer (2.5) by refering to a canonical probability space $(\Omega, \mathcal{A}, \mathbb{P})$ into a definition of the Palm distribution in the same spirit as (2.20). The original formulation of the following theorem can be found in [75] (cmp. also [53]).

Theorem 2.4 (Neveu's exchange formula) *Let X_D and $\widetilde{X}_{\widetilde{D}}$ be two arbitrary jointly stationary marked point processes in $\mathbb{R}^2$ with mark spaces $\mathbb{M}$ and $\widetilde{\mathbb{M}}$ and intensities λ and $\widetilde{\lambda}$, respectively. Then, for any measurable function $f : \mathbb{R}^2 \times \mathbb{M} \times \widetilde{\mathbb{M}} \times \Omega \to [0, \infty)$,*

$$\lambda \int_{\Omega \times \mathbb{M}} \int_{\mathbb{R}^2 \times \widetilde{\mathbb{M}}} f(x, g, \widetilde{g}, \theta_x \omega) \widetilde{X}_{\widetilde{D}}(\omega, d(x, \widetilde{g})) P^*_{X_D}(d(\omega, g))$$

$$= \widetilde{\lambda} \int_{\Omega \times \widetilde{\mathbb{M}}} \int_{\mathbb{R}^2 \times \mathbb{M}} f(-x, g, \widetilde{g}, \omega) X_D(\omega, d(x, g)) P^*_{\widetilde{X}_{\widetilde{D}}}(d(\omega, \widetilde{g})) \,. \tag{2.21}$$

2.3 The Boolean Model and Modulated Poisson Point Processes

A relatively simple but very important model for stationary random closed sets is represented by the Boolean model. This model is a specification of a germ–grain model where the germs are formed by a Poisson point process and where the grains are generated by a sequence of independent and identically distributed random closed sets. In recent years the Boolean model has been used in various applications, for example, in material sciences, telecommunication, and biology. A detailed study of it can be found in [55] and [96]. In [70] and [100] overviews are given. In this thesis Boolean models are mainly used in order to define modulated Poisson point processes that are modulated by Boolean models (see Section 2.3.2).

2.3.1 Definition of the Boolean Model

The Boolean model can be considered as a special case of a *germ–grain model* that is defined as a marked point process $Y_D = \{Y_n, M_n\}$ on $\mathbb{R}^2$ if for the corresponding mark space we have that $(\mathbb{M}, \mathcal{M}) = (\mathcal{F}, \mathcal{B}(\mathcal{F}))$. The point process $Y_D = \{Y_n, M_n\}$ is called *germ–grain process*, where the sequence $\{Y_n\}_{n \geq 1}$ is called the *germ process* and the sequence $\{M_n\}_{n \geq 1}$ is called the *grain process*. Often, M_0 is called the *primary grain* and is a representant of the sequence $\{M_n\}_{n \geq 1}$ in the sense that $M_1, M_2, \ldots$ are independent of each other and identically distributed with respect to M_0.

For a formal definition of the Boolean model, regard a stationary Poisson point process $Y = \{Y_n\}_{n \geq 1}$ in $\mathbb{R}^2$ with intensity $\beta > 0$. Let $\{M_n\}_{n \geq 1}$ denote a sequence of independent and identically distributed random compact sets in $\mathbb{R}^2$, where M_0 is a random compact set that has the same distribution as M_n. For any compact set $K \subset \mathbb{R}^2$ let

$$\mathbb{E}(\nu_2(M_0 \oplus K)) < \infty,$$

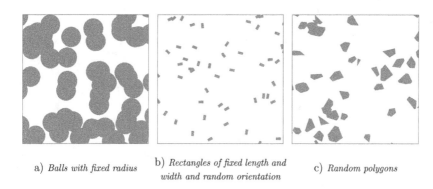

a) *Balls with fixed radius*   b) *Rectangles of fixed length and width and random orientation*   c) *Random polygons*

Figure 2.4: Realisations of Boolean models with different grains

where $\oplus$ denotes Minkowsky addition (see A.1). Additionally, let the sequences $\{Y_n\}_{n\geq 1}$ and $\{M_n\}_{n\geq 1}$ be independent. Then the random closed set Ψ given by

$$\Psi = \bigcup_{n\geq 1}(M_n + Y_n)$$

is called a *Boolean germ–grain model.*

Note that due to the fact that the Poisson point process $Y = \{Y_n\}_{n\geq 1}$ is assumed to be stationary and independent of the sequence of marks $\{M_n\}_{n\geq 1}$ we can consider the germ–grain process $Y_D = \{Y_n, M_n\}$ as a stationary Poisson point process in $\mathbb{R}^2$ with independent marks. It can be shown that the Boolean model is indeed a stationary random closed set (cmp. [94], p. 100).

The coverage probability p_Ψ for a Boolean model can be computed explicitly as ([100], pp. 64ff.)

$$p_\Psi = \mathbb{P}(o \in \Psi) = 1 - \exp\left(-\beta\mathbb{E}\nu_2(M_0)\right), \qquad (2.22)$$

where β is the intensity of the underlying stationary Poisson point process Y and M_0 represents the primary grain.

In the following we will focus on Boolean models with circular grains of a fixed or a random but bounded radius. But, in general, there are of course numerous possibilities for the choice of the grains. Figure 2.4 illustrates some of these possibilities by displaying realisations of Boolean models where the grains are given by balls with a fixed radius, by rectangles of fixed length and width and random orientation, and by random polygons, respectively.

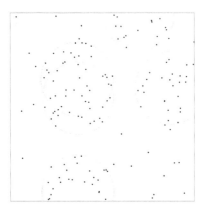

Figure 2.5: Realisation of a modulated Poisson point processes with corresponding realisation of the Boolean model shown in red

2.3.2 Modulated Poisson Point Processes Based on Boolean Models

Based on the Boolean model defined in Section 2.3.1 we are able to define modulated Poisson point processes that are an extension of stationary Poisson point processes and can be seen as a special type of Cox point processes introduced in Section 2.2.5.

Let Ψ be a Boolean model with finite intensity $\beta > 0$ and circular grains with a fixed radius r. Furthermore, let X be a (planar) Cox point process that has a random driving measure Λ_X which is defined by

$$\Lambda_X(dx) = \left\{ \begin{array}{ll} \lambda_{X_1} dx & \text{if}\quad x \in \Psi, \\ \lambda_{X_2} dx & \text{if}\quad x \notin \Psi, \end{array} \right. \tag{2.23}$$

where $0 \leq \lambda_{X_i} < \infty$ for $i \in \{1, 2\}$ and $\max\{\lambda_{X_1}, \lambda_{X_2}\} > 0$. Then we call the Cox point process X a *modulated Poisson point process*. We often refer to the Boolean model Ψ as the *corresponding* or *underlying* Boolean model with respect to X. Obviously, the modulated Poisson point process X with random driving measure Λ_X given in (2.23) can completely be described by the four parameters $p_\Psi, \beta, \lambda_{X_1}, \lambda_{X_2}$ since the (fixed) radius r of the Boolean model Ψ can be computed, given β and p_Ψ, by using (2.22). In Figure 2.5 a realisation of a modulated Poisson point process is displayed. Note that an extension to the case where the grains of Ψ have a random but bounded radius is possible without any problems.

Lemma 2.6 *Let X be a modulated Poisson point process with random driving measure Λ_X given in (2.23). Then X is stationary with intensity $\lambda_X = p_\Psi \lambda_{X_1} + (1 - p_\Psi)\lambda_{X_2}$, where p_Ψ is given in (2.22).*

Proof From the stationarity of the Boolean model Ψ and the definition of the modulated Poisson point process X the stationarity of X can directly be deduced using Lemma 2.5. Let $B \in \mathcal{B}_0(\mathbb{R}^2)$ be an arbitrary bounded Borel set. It holds that

$$\mathbb{E}X(B) = \mathbb{E}X(B \cap \Psi) + \mathbb{E}X(B \cap \Psi^c)$$

Using the definition of Λ_X given in (2.23) and the stationarity of Ψ we obtain that

$$\mathbb{E}X(B \cap \Psi) = \lambda_{X_1}\mathbb{E}\left[\nu_2(B \cap \Psi)\right] = p_\Psi \lambda_{X_1} \nu_2(B)$$

and that

$$\mathbb{E}X(B \cap \Psi^c) = \lambda_{X_2}\mathbb{E}\left[\nu_2(B \cap \Psi^c)\right] = (1 - p_\Psi)\lambda_{X_2}\nu_2(B).$$

Altogether we get that

$$\lambda_X \nu_2(B) = (p_\Psi \lambda_{X_1} + (1 - p_\Psi)\lambda_{X_2})\nu_2(B).$$

Since B is chosen arbitrarily in $\mathcal{B}_0(\mathbb{R}^2)$ the proof is completed. $\square$

Another important property of the modulated Poisson point process X is provided by the lemma hereafter.

Lemma 2.7 *Let X be a modulated Poisson point process with random driving measure Λ_X given in (2.23). Then X is ergodic.*

A proof of this lemma can be given analogously to the proof of Lemma 2.4 using the semiring defined in (2.4). In particular we have for two sets $B_1, B_2 \in \mathcal{B}_0(\mathbb{R}^2)$ that the random numbers of points $X(B_1)$ and $X(B_2)$ are independent if $\min\{|x - y| : x \in B_1, y \in B_2\} > 2r$, where r is the (fixed) radius of the grains of the Boolean model Ψ.

Note that although the modulated Poisson point process is stationary, it is a very flexible model that is suitable to model a large variety of scenarios. It is possible to define modulated Poisson point processes in a more general setting using random or deterministic partitions of the plane (cmp. [15], [16]). With regard to this thesis we will focus on modulated Poisson point processes with respect to a Boolean model with circular grains of a fixed or random but bounded radius.

2.4 Definition of Random Tessellations with Examples

In this section we turn our attention to the introduction of random tessellations (also called random mosaics), the particular model of stochastic geometry that is common to all applications regarded in this thesis. After the definition of a deterministic tessellation we will clarify what is meant by a random tessellation. Some examples for random tessellations are shown that are used during the following chapters either for model fitting of network structures or for cost analysis. Among them are the Poisson–Voronoi tessellation, the Poisson–Delaunay tessellation, the Poisson line (or more general hyperplane) tessellation as well as more sophisticated models based on these basic tessellation models like modulated and iterated tessellations.

2.4.1 Deterministic Tessellations

Consider $\mathcal{P}$ as the family of all compact and convex polygons $P \subset \mathbb{R}^2$ with int $P \neq \emptyset$, where int P denotes the interior of P. Recall that P is called a (convex) *polygon* if it can be represented as the convex hull of a point set $\{x_1, ..., x_n\} \in \mathbb{R}^2$. Now regard $T = \{P_n\}_{n \geq 1} \subset \mathcal{P}$ as a sequence of polygons. The sequence T is said to be a (deterministic) *tessellation* if

- int $P_i \cap$ int $P_j = \emptyset$, $\forall P_i, P_j \in T$ with $P_i \neq P_j$, i.e., $\{P_n\}_{n \geq 1}$ are pairwise non–overlapping,

- $\bigcup_{P_i \in T} P_i = \mathbb{R}^2$, i.e., the union of polygons covers the plane completely, and

- for any compact set $C \in \mathcal{C}$ the following statement holds
  $T(C) = \{P \in T : P \cap C \neq \emptyset\}$ is finite, i.e., T is a locally finite family.

2.4.2 Definition of Random Tessellations

Based on the definition of deterministic tessellations we are able to define a *random tessellation* in the following way.

Let $\mathbb{T}$ denote the class of all tessellations in $\mathbb{R}^2$. A sequence $\tau = \{\Xi_n\}_{n \geq 1}$ of random convex polygons is called a *random tessellation* if

$$\mathbb{P}\left(\{\Xi_n\}_{n \geq 1} \in \mathbb{T}\right) = 1.$$

With respect to random tessellations we can define stationarity and isotropy analogously to the definitions for random (marked) point processes, meaning that a random tessellation τ is stationary if its distribution P_τ is invariant under translation in $\mathbb{R}^2$ and it is isotropic if its distribution P_τ is invariant under rotation around the origin. If a random tessellation is both stationary and isotropic it is called *motion–invariant*.

For many applications it is often very convenient to represent a (stationary) random tessellation $\tau = \{\Xi_n\}_{n\geq 1}$ as a (stationary) random marked point process in $\mathbb{R}^2$ by introducing the concept of *associated points* (see e.g Chapter 4 and in particular Satz 4.3.1 in [94]).

Let $\alpha : \mathcal{K} \to \mathbb{R}^2$ be a measurable mapping that satisfies

$$\alpha(K) \in \mathcal{K} \text{ and } \alpha(K + x) = \alpha(K) + x \qquad (2.24)$$

for any $K \in \mathcal{K}$ and $x \in \mathbb{R}^2$. For a stationary random tessellation $\tau = \{\Xi_n\}_{n\geq 1}$ we call $\alpha(\Xi_n)$ the *associated point* of the nth cell Ξ_n, $n \geq 1$. Some possible choices of the associated point are the lexicographically smallest vertex or the nucleus of the cell. Using the mapping α described in 2.24 we are able to express a stationary random tessellation $\tau = \{\Xi_n\}_{n\geq 1}$ as a marked point process $\tau_X = \sum_{n\geq 1} \delta_{(\alpha(\Xi_n),\Xi_n^o)}$, where $\Xi_n^o = \Xi_n - \alpha(\Xi_n)$ is called the *nth centered cell* containing the origin o.

In the course of this thesis both ways of expressing random tessellations are used, either as a sequence of random convex polygons denoted by τ or as a marked point process, where the marks are given by the random convex polygons and the locations are given by the associated points with respect to the marks denoted as τ_X. Note that this is not really a difference since we are always able to transfer τ into τ_X and vice versa by defining suitable associated points for each random polygon and by looking at the sequence of random polygons that is induced by τ_X, respectively.

Suppose that the random tessellation $\tau_X = \sum_{n\geq 1} \delta_{(\alpha(\Xi_n),\Xi_n^o)}$ is stationary with positive and finite intensity $\lambda_\tau = \mathbb{E}\#\{n : \alpha(\Xi_n) \in [0,1)^2\}$. Denote by $\mathcal{P}^o$ the family of all convex polytopes with their associated point at the origin. The *Palm mark distribution* $P_{\tau_X}^*$ of τ_X is given by

$$P_{\tau_X}^*(B) = \lambda_\tau^{-1}\, \mathbb{E}\#\{n : \alpha(\Xi_n) \in [0,1)^2,\ \Xi_n^o \in B\}$$

for any $B \in \mathcal{B}(\mathcal{F}) \cap \mathcal{P}^o$. Notice that a random polytope $\Xi_\tau^* : \Omega \to \mathcal{P}^o$, whose distribution coincides with $P_{\tau_X}^*$, is called the *typical cell* of the tessellation τ_X. Hence, we can think of the typical cell as a cell that is drawn uniformly with respect to the pool of all cells available for the random tessellation τ_X.

With regard to the expected area $\mathbb{E}\nu_2(\Xi_\tau^*)$ of the typical cell Ξ_τ^* we obtain the following result which states that the expected area of the typical cell is reciprocal to the intensity λ_τ of the random tessellation τ_X.

Lemma 2.8 *Let τ_X be a stationary random tessellation with positive and finite intensity λ_τ and let Ξ_τ^* denote the typical cell of τ_X. Then it holds that*

$$\mathbb{E}\nu_2(\Xi_\tau^*) = \frac{1}{\lambda_\tau} = \int_{\mathcal{P}^o} \nu_2(C)\, P_{\tau_X}^*(dC). \qquad (2.25)$$

A proof for this lemma can be found in [95] on pp. 237f.

The cell Ξ_τ^o of a stationary random tessellation $\tau = \{\Xi_n\}_{n\geq 1}$ that contains the origin o is often denoted as the *zero cell*, i.e.,

$$\Xi_\tau^o = \Xi_k \ \text{ if } \ o \in \text{int } \Xi_k.$$

With respect to the typical cell Ξ_τ^* and the zero cell Ξ_τ^o of a stationary random tessellation τ_X we have the following relationship.

Lemma 2.9 *Let τ_X be a stationary tessellation and let Ξ_τ^* and Ξ_τ^o be the corresponding typical cell and zero cell, respectively. Additionally, let $f : \mathcal{C} \to [0,\infty)$ be an arbitrary, translation–invariant, non–negative and measurable function. Then*

$$\mathbb{E}\left[f(\Xi_\tau^o)\right] = \lambda_\tau \mathbb{E}\left[f(\Xi_\tau^*)\nu_2(\Xi_\tau^*)\right]. \qquad (2.26)$$

A proof for this lemma using the Campbell Theorem can be found in [95] on pp. 252f. We can deduce from (2.26) that the distribution of the functional $f(\Xi_\tau^o)$ of the zero cell Ξ_τ^o can be considered as the area weighted distribution of the functional $f(\Xi_\tau^*)$ for the typical cell Ξ_τ^*.

2.4.3   Poisson Line Tessellations

As a first example for a random tessellation we regard Poisson line tessellations. In order to define them we have to define a line process first.

Let $\mathcal{S}$ be the set of all one–dimensional subspaces of $\mathbb{R}^2$ and let $\mathcal{L} = \{L \in \mathcal{S} : o \in L\}$. A point process X_ℓ in $\mathcal{F}' = \mathcal{F} \setminus \{\emptyset\}$ is called a (planar) *line process* if for the intensity measure Λ_{X_ℓ} of X_ℓ it holds that

$$\Lambda_{X_\ell}(\mathcal{F}' \setminus \mathcal{S}) = 0.$$

We can define stationarity, isotropy and motion–invariance of line processes analogously to the case of point processes. If the line process X_ℓ is stationary we can disintegrate Λ_{X_ℓ} as follows. Suppose that Λ_{X_ℓ} is locally finite and not equal to the zero measure.

There exists a constant $\lambda_{X_\ell} \in (0, \infty)$ and a probability measure θ on $\mathcal{B}(\mathcal{L})$, called the *orientation distribution* of X_ℓ, such that

$$\Lambda_{X_\ell}(B) = \lambda_{X_\ell} \int_{\mathcal{L}} \int_{L^\perp} \mathbb{I}_B(L + x)\nu_1(dx)\theta(dL) \qquad (2.27)$$

for any $B \in \mathcal{B}(\mathcal{S})$, where ν_1 denotes the one–dimensional Lebesgue measure on the orthogonal complement $L^\perp \in \mathcal{L}$ of $L \in \mathcal{L}$. Note that (2.27) yields that

$$\lambda_{X_\ell} = \frac{1}{2} \mathbb{E}X_\ell(L \in \mathcal{S} : L \cap b(o, 1) \neq \emptyset) , \qquad (2.28)$$

i.e., $2\lambda_{X_\ell}$ is the expected number of lines hitting $b(o, 1)$, the disc with center o and unit radius.

In the following, we will consider the case that X_ℓ is a stationary and isotropic *Poisson line process*. Then, X_ℓ can be represented in the form $X_\ell = \sum_{n \geq 1} \delta_{\ell_{(R_n, V_n)}}$, where $\{R_n\}$ is a stationary Poisson point process in $\mathbb{R}_+$ with intensity $2\lambda_{X_\ell}$ and $\{V_n\}$ is an independent sequence of independent and identically distributed random variables with uniform distribution on $[0, 2\pi)$. For each line $\ell_{(R_n, V_n)}$, the angle V_n is measured in anti–clockwise direction between the (positive) x–axis and the outer orientation vector of the line, whereas R_n denotes the perpendicular distance of the line to the origin. Note that, for a stationary isotropic line process, (2.27) can be written as

$$\Lambda_{X_\ell}(B) = \frac{\lambda_\ell}{2\pi} \int_0^{2\pi} \int_0^\infty \mathbb{I}_B(\ell_{(r,v)})dr dv, \qquad B \in \mathcal{B}(\mathcal{S}) . \qquad (2.29)$$

Recall that each line $\ell_{(R_n, V_n)}$ in $\mathbb{R}^2$ can be described by its Hessian normal form $\ell_{(R_n, V_n)} = \{(x, y) \in \mathbb{R}^2 : x \cos V_n + y \sin V_n = R_n\}$. It can be shown that the expected total length $\mathbb{E}\sum_{n \geq 1} \nu_1(\ell_{(R_n, V_n)} \cap b(o, 1))$ of lines $\ell_{(R_n, V_n)}$ in $b(o, 1)$ is given by $\pi\lambda_{X_\ell}$ (cmp. e.g. p. 324 in [100]). Thus, $\gamma = \lambda_{X_\ell}$ is the expected total length of the lines per unit area and is called the *intensity* of the random closed set $X_\ell = \bigcup_{n \geq 1} \ell_{(R_n, V_n)}$. In the following, we will focus on isotropic Poisson line processes but keep in mind that other choices with respect to the orientation distribution of the lines are also possible.

From the definition of Poisson line processes a definition of Poisson line tessellations is derived in a canonical way. Let X_ℓ be a Poisson line process. The random tessellation τ_{X_ℓ} induced by X_ℓ is called a *Poisson line tessellation* (PLT). Figure 2.6 shows a realisation of an isotropic Poisson line tessellation, whereas Figure 2.7 shows two realisations for anisotropic cases. In Figure 2.7a we can see the so–called *Manhattan model*, where the angles of the lines with respect to the x–axis are either 0 or $\pi/2$. In Figure 2.7b the angles of the lines with respect to the x–axis are uniformly distributed on the interval $[0, 1]$. Note that, for dimensions greater than two, Poisson line tessellations are called

Figure 2.6: Realisation of an isotropic Poisson line tessellation

Poisson hyperplane tessellations. An isotropic Poisson line tessellation τ_{X_ℓ} can be described completely by the parameter γ which is the intensity of the Poisson line process X_ℓ that induces τ_{X_ℓ}. Recall that γ in this context represents the mean total length of the lines of X_ℓ per unit area.

2.4.4 Poisson–Voronoi Tessellations

One of the most famous examples for a random tessellation is given by the Poisson–Voronoi tessellation. A Voronoi tessellation is based on a given locally finite set of points $B \subset \mathbb{R}^2$. It uses the nearest neighbour principle. A point belongs to a certain cell containing a so-called nucleus $x \in B$ if its nearest neighbour with respect to B is x. A very detailed discussion of Voronoi tessellations in general can be found in [78]. Formally, a (deterministic) *Voronoi tessellation* can be defined as follows.

Let $B = \{x_1, x_2, ...\}$ be a locally finite set of points in $\mathbb{R}^2$ whose convex hull $conv B$ is the whole Euclidean plane $\mathbb{R}^2$. For $x_n, x_m \in B$ define the halfplane $H(x_n, x_m)$ by

$$H(x_n, x_m) = \{x \in \mathbb{R}^2 : |x - x_n| \le |x - x_m|\}.$$

Then we call the polygon $P(x_n)$ given by

$$P(x_n) = \bigcap_{m \neq n} H(p_n, p_m) = \{x \in \mathbb{R}^2 : |x - x_n| \le |x - x_n|, \ \forall \ m \neq n\} \qquad (2.30)$$

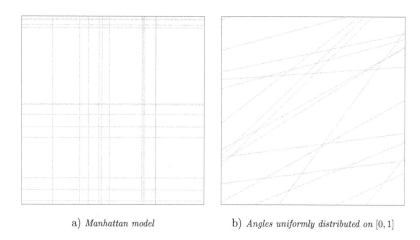

a) *Manhattan model*   b) *Angles uniformly distributed on* $[0,1]$

Figure 2.7: Realisations of anisotropic Poisson line tessellations

the *Voronoi cell* of the point x_n. The sequence $\tau = \{P(x_n) : x_n \in B\}$ is called the *Voronoi tessellation* with respect to B. The point x_n is often denoted as the *nucleus* of $P(x_n)$. Note that due to the assumption given above that $convB = \mathbb{R}^2$, i.e. that the convex hull of B is the whole Euclidean plane we obtain that the polygons $P(x_n)$ are bounded for any $n \geq 1$ (cmp. p. 256 in [94]).

After the definition of a deterministic Voronoi tessellation the *Poisson–Voronoi tessellation* (PVT) can be defined in a very natural fashion as the Voronoi tessellation τ_X that is induced by a Poisson point process $X = \{X_1, X_2, ...\}$, i.e., where X acts as the (random) set of nuclei. Therefore, a Poisson–Voronoi tessellation can be derived by a two step mechanism. First we have a Poisson point process $X = \{X_1, X_2, ...\}$ and in the second step a Voronoi tessellation τ_X is constructed based on $X = \{X_1, X_2, ...\}$. A realisation of a stationary Poisson–Voronoi tessellation is shown in Figure 2.8a, while a realisation of an instationary Poisson–Voronoi tessellation is displayed in Figure 2.8b. Due to the definition we are able to describe a stationary Poisson–Voronoi tessellation τ_X by a single parameter γ that reflects the intensity of the generating stationary Poisson point process that is the same as the expected number of nuclei in a window of unit area.

For a Voronoi tessellation τ_X that is induced by a random point process X (not necessarily Poisson), i.e., whose nuclei are generated by a random point process, it is easy to see that stationarity, isotropy and motion–invariance of τ_X directly follow from the

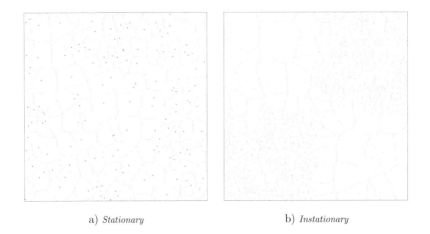

a) *Stationary*   b) *Instationary*

Figure 2.8: Realisation of Poisson–Voronoi tessellations

same properties for X, since this is a direct consequence of the construction principle. The same relationship holds with respect to ergodicity.

Note that it is possible to define a Voronoi tessellation not only with respect to a locally finite set B of points in $\mathbb{R}^2$ but to more general sets. For example, we can regard a (locally finite) set B' of lines in $\mathbb{R}^2$ and construct the Voronoi tessellation τ' with respect to B'. In this case, each line segment that is the result of the lines intersecting each other forms a Voronoi cell by using the nearest neighbour principle described in (2.30). Analogously to the case of a Poisson–Voronoi tessellation that is induced by a Poisson point process we are also able to define a Voronoi tessellation that is induced by the segments of a Poisson line process introduced in Section 2.4.3. A realisation of such a Voronoi–tessellation with respect to the lines of a Poisson line process is displayed in Figure 2.9.

2.4.5 Poisson–Delaunay Tessellations

A tessellation model that is very closely related to Poisson–Voronoi tessellations are Poisson–Delaunay tessellations or more precisely Poisson–Delaunay triangulations. A Delaunay tessellation can be considered as the dual graph to the Voronoi tessellation in the sense that there is a one–to–one relationship between the vertices of the Delaunay tessellation and the cells of the corresponding Voronoi tessellation.

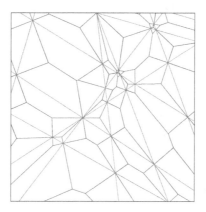

Figure 2.9: Realisation of a Voronoi tessellation (blue) induced by the segments of a Poisson line process (red)

Let $B \subset \mathbb{R}^2$ be a set of points that is not collinear, i.e., if x_i, x_j, x_k are three pairwise different points in B it holds that there does not exist a line with the property that x_i, x_j, x_k are all located on that line. Furtermore, let $\tau' = \{P(x_n)\}$ be the Voronoi tessellation with respect to B. Let $Q = \{q_1, q_2, ...\}$ be the set of vertices of τ' and $x_{i_1}, ..., x_{i_{k_i}}$ be the points in B whose Voronoi cells share the vertex q_i. Let

$$T_i = \{x \in \mathbb{R}^2 : x = \sum_{j=1}^{k_i} \lambda_j x_{i_j}, \quad \sum_{j=1}^{k_i} \lambda_j = 1, \quad \lambda_j \geq 0\}$$

and let $\tau = \{T_1, T_2, ...\}$. If $k_i = 3$ for all i the set τ is called the *Delaunay tessellation* with respect to B.

Note that if $k_i > 3$ for at least one i, the tessellation τ is called a Delaunay pretriangulation. Otherwise we call it a Delaunay triangulation. The term triangulation here means that every polygon of the tessellation is a triangle. It is easily possible to construct a triangulation out of a pretriangulation.

The definition of the Poisson–Delaunay triangulation is achieved similarily to the definition of the Poisson–Voronoi tessellation. Let $X = \{X_1, X_2, ...\}$ be a stationary Poisson point process and let τ'_X be the Poisson–Voronoi tessellation induced by X. Then, the corresponding Delaunay tessellation τ_X that is induced by X and τ'_X is called a *Poisson–Delaunay tessellation* (PDT). Hence, similarly as for a Poisson–Voronoi tessellation, we can think of a Poisson–Delaunay tessellation as a two step mechanism. First we have

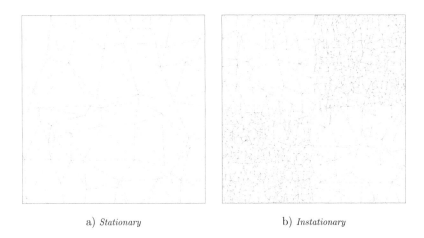

a) *Stationary*   b) *Instationary*

Figure 2.10: Realisations of Poisson–Delaunay tessellations

a Poisson point process $X = \{X_1, X_2, ...\}$ and in a second step the Delaunay tessellation is derived that is based on X. A realisation of a stationary Poisson–Delaunay tessellation is displayed in Figure 2.10a, while a realisation of an instationary Poisson–Delaunay tessellation is shown in Figure 2.10b. Note that, similarly to the Poisson–Voronoi tessellation, the stationary Poisson–Delaunay tessellation can be described by a single parameter γ that reflects the intensity of the generating stationary Poisson point process. Hence, this parameter describes the expected number of vertices of the Poisson–Delaunay tessellation per unit area.

As mentioned in the introduction, the Delaunay tessellation can be considered as a dual graph to a corresponding Voronoi tesssellation in the sense that

1. each triangle of the Delaunay tessellation corresponds to a vertex of the Voronoi tessellation,

2. each edge of the Delaunay tessellation corresponds to an edge of the Voronoi tessellation,

3. each vertex of the Delaunay tessellation corresponds to a cell of the Voronoi tessellation.

2.4.6 Cox–Voronoi Tessellations Based on a Poisson Line Process

Using the concept of doubly stochastic Poisson or Cox point processes introduced in Section 2.2.5 and Poisson line processes introduced in Section 2.4.3, we are able to define a Cox point process in $\mathbb{R}^2$ that is located on the (random) lines of a Poisson line process. More precisely, let X_ℓ be a stationary and isotropic Poisson line process with intensity γ. Then, define the Cox process X_c as a doubly stochastic Poisson point process with random driving measure Λ_{X_c} given by

$$\Lambda_{X_c}(B) = \lambda_\ell \nu_1(B \cap X_\ell) \tag{2.31}$$

for some $\lambda_\ell > 0$, $B \in \mathcal{B}(\mathbb{R}^2)$. The definition of the random driving measure Λ_{X_c} in (2.31) induces that a realisation of X_c can be obtained by first realising a Poisson line process X_ℓ and afterwards realising (linear and stationary) Poisson point processes with intensity λ_ℓ on the lines of the realisation of X_ℓ.

Lemma 2.10 *Let X_c be a Cox point process with random driving measure Λ_{X_c} defined in (2.31). Then X_c is stationary and isotropic and the intensity λ_{X_c} of X_c is given by $\lambda_{X_c} = \lambda_\ell \gamma$.*

Proof The stationarity and isotropy of X_c follow directly from the stationarity and isotropy of X_ℓ and from the Poissonian placement of the points of X_c on X_ℓ. The random driving measure Λ_{X_c} of X_c given in (2.31) yields for any $B \in \mathcal{B}_0(\mathbb{R}^2)$ the relationship

$$\Lambda_{X_c}(B) = \mathbb{E} X_c(B) = \lambda_\ell \mathbb{E} \nu_1(B \cap X_\ell) = \lambda_\ell \Lambda_{X_\ell}(B),$$

where $\Lambda_{X_\ell}(B)$ is the intensity measure of X_ℓ. For a stationary and isotropic Poisson line process X_ℓ it holds that

$$\Lambda_{X_\ell}(B) = \mathbb{E} X_\ell(B) = \gamma \nu_2(B).$$

Hence, we obtain that

$$\Lambda_{X_c}(B) = \lambda_\ell \gamma \nu_2(B),$$

for arbitrary $B \in \mathcal{B}_0(\mathbb{R}^2)$. Therefore, X_c has intensity $\lambda_{X_c} = \lambda_\ell \gamma$. □

Note that if we regard X_c with respect to a specific line of X_ℓ, it represents a one–dimensional Poisson point process with intensity λ_ℓ. Hence, we can interpret λ_ℓ as the mean number of points per unit length of X_ℓ. In Figure 2.11b, a realisation of a Cox point process X_c with random driving measure Λ_{X_c} defined in (2.31) is shown which is based on a realisation of a Poisson line process X_ℓ displayed in Figure 2.11a. A

Voronoi tessellation τ_{X_c} that is induced by the Cox point process X_c can be obtained in the canonical way by using the points of X_c as the (random) nuclei of τ_{X_c} and by applying the nearest neighbour principle given in (2.30). In Figures 2.11c and 2.11d it is visualized how such a Cox–Voronoi tessellation (CVT) is obtained.

a) *Realisation of a Poisson line process X_ℓ*   b) *Realisation of points of X_c on the lines of X_ℓ*

c) *Voronoi cells having points of X_c as nuclei*   d) *Realisation of a Cox–Voronoi tessellation τ_{X_c}*

Figure 2.11: Construction principle for the Cox–Voronoi tessellation τ_{X_c}

Scaling property
With respect to the Cox–Voronoi tessellation τ_{X_c} that is induced by a Cox point process X_c with random driving measure Λ_{X_c} given in (2.31) an important scaling property can be stated. Scaling in this context means that for specific parameter configurations identical random structures can be observed but on different length scales. More precisely, consider the parameters of the Cox–Voronoi tessellation given as γ, the parameter of the Poisson line process X_ℓ, and as λ_ℓ, the average number of points of X_c per unit length of X_ℓ, respectively. Then, if we introduce the scaling parameter $\kappa = \gamma/\lambda_\ell$, a

scaling effect is realised for this specific tessellation model (Figure 2.12). Especially for characteristics of the typical cell Ξ_τ^* of the Cox–Voronoi tessellation τ_{X_c} this is a useful property.

For example, assuming $\kappa = \gamma/\lambda_\ell$ to be constant, the mean perimeter and the square root of the mean area of the typical cell behave linearly with respect to $1/\gamma$ (the mean edge length for the Poisson line tessellation). This is due to the fact that for constant κ we are realising identical random structures but on different length scales, where a suitable measurement for the length scale is the mean edge length of the underlying Poisson line tessellation. Hence we have that the product of mean perimeter of the typical cell times γ is constant as well as the product of the square root of the mean area of the typical cell times γ as long as κ is fixed. Note that, with respect to the scaling invariance of the square root of the mean area of the typical cell, it is also possible to provide a proof using Lemmas 2.6 and 2.8.

In this sense a scaling invariance property is given which can be used to partition the originally two–dimensional parameter space into equivalence classes and to thereby reduce in a certain sense the number of parameters from two, γ and λ_ℓ, to a single one $\kappa = \gamma/\lambda_\ell$. Additionally, this property provides us with a suitable criterion for the development of software tests which will be applied in Chapter 5.

Figure 2.12: Scaling effect, same random structures but on different scales

2.4.7 Modulated Tessellations

Based on the Voronoi tessellation and the Delaunay tessellation introduced in Sections 2.4.4 and 2.4.5, respectively, and based on the modulated Poisson point process

X with random driving measure Λ_X introduced in Section 2.3.2 we are able to construct more sophisticated tessellation models that are useful for applications like in the field of telecommunication.

Let X be a modulated Poisson point process with random driving measure Λ_X given in (2.23). Then we call the (random) Voronoi tessellation τ_X that is induced by X a *modulated Poisson–Voronoi tessellation* and, analogously, we call the corresponding Delaunay tessellation τ'_X which is also induced by X a *modulated Poisson–Delaunay tessellation*. A realisation of a modulated Poisson–Voronoi tessellation and a realisation of a modulated Poisson–Delaunay tessellation are displayed in Figures 2.13a and 2.13b, respectively. Two special cases of modulated tessellations are given if either $\lambda_{X_1} = 0$ or if $\lambda_{X_2} = 0$. If $\lambda_{X_1} = 0$ we speak about a *Swiss cheese model* (Figure 2.14a), whereas the case $\lambda_{X_2} = 0$ is denoted as an *Inner–city model* (Figure 2.14b).

Scaling property
Similarly to the scaling property described in Section 2.4.6 for the case of a Voronoi tessellation τ_{X_c} induced by a Cox point process X_c, we are able to provide a scaling invariance effect for a Voronoi tessellation τ_X that is induced by a modulated Poisson point process X. Recall that a modulated Poisson point process X as well as the Voronoi tessellation τ_X induced by X can be completely described by a vector of four parameters $(p_\Psi, \beta, \lambda_{X_1}, \lambda_{X_2})$, where β is the intensity of the germs of the Boolean model Ψ, p_Ψ is the coverage probability of Ψ given in (2.22) and λ_{X_1} and λ_{X_2} are the two intensities appearing in the definition of the random driving measure Λ_X of X given in (2.23). Now regard the vector $\kappa' = (p_\Psi, \lambda_{X_1}/\beta, \lambda_{X_2}/\beta)$.

Then it holds for two different Voronoi tessellations $\tau_X^{(1)}$ and $\tau_X^{(2)}$ that are induced by two modulated Poisson point processes $X^{(1)}$ and $X^{(2)}$ with corresponding parameter vectors $(p_\Psi^{(1)}, \beta^{(1)}, \lambda_{X_1}^{(1)}, \lambda_{X_2}^{(1)})$ and $(p_\Psi^{(2)}, \beta^{(2)}, \lambda_{X_1}^{(2)}, \lambda_{X_2}^{(2)})$ that a scaling invariance effect is realised if $\kappa'^{(1)} = \kappa'^{(2)}$. Scaling invariance in this case means that, with respect to the typical cell Ξ_τ^* of τ_X, it holds that

$$\mathbb{E}\left[\nu_1(\partial\Xi_\tau^{*(1)})\right]\sqrt{\beta^{(1)}} = \mathbb{E}\left[\nu_1(\partial\Xi_\tau^{*(2)})\right]\sqrt{\beta^{(2)}}, \qquad (2.32)$$

and that

$$\sqrt{\mathbb{E}\left[\nu_2(\Xi_\tau^{*(1)})\right]}\sqrt{\beta^{(1)}} = \sqrt{\mathbb{E}\left[\nu_2(\Xi_\tau^{*(2)})\right]}\sqrt{\beta^{(2)}}, \qquad (2.33)$$

if $\kappa'^{(1)} = \kappa'^{(2)}$. Note that, with regard to the mean number of vertices for the typical cell, no rescaling has to be performed since it is always equal to 6 and independent of the parameters used. The relationships (2.32) and (2.33) for the Voronoi tessellation induced by a modulated Poisson point process X are very similar to the relationships given in Section 2.4.6 for the Voronoi tessellation induced by a Cox point process X_c. Once more, we obtain in (2.32) that the mean perimeter times a suitable length scale is constant

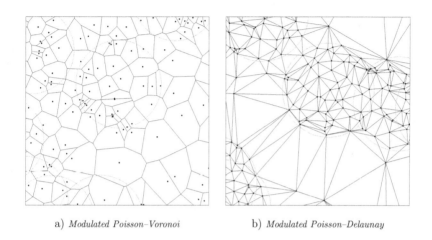

a) *Modulated Poisson–Voronoi*   b) *Modulated Poisson–Delaunay*

Figure 2.13: Realisations of modulated Poisson–type tessellations with realisations of corresponding Boolean models in red

for different parameter configurations asuming that a scaling parameter (vector) is kept constant. A similar relationship for the square root of the mean area is given in (2.33) which could in this case also be proven using Lemmas 2.6 and 2.8. An implication of this scaling invariance property is that it suffices to regard a three–dimensional parameter space $(p_\Psi, \lambda_{X_1}/\beta, \lambda_{X_2}/\beta)$ instead of the originally four–dimensional parameter space $(p_\Psi, \beta, \lambda_{X_1}, \lambda_{X_2})$. Thereby a systematic analysis is of course enormously facilitated.

2.4.8   Iterated Tessellations

With regard to applications there is often a need to have a repertoire of possible models that are more flexible than the tessellation models introduced in Sections 2.4.4–2.4.7. Such quite basic models often do not reflect the given data to a satisfactory extend. Thus, these tessellation models must be extended to more sophisticated tessellation models that are able to cope with real data structures. On the other hand it is obvious that these models should still be tractable at least from a computational point of view. Examples for construction principles that lead to more flexible and sophisticated tessellation models are the superposition and the nesting of random tessellations or more generally the iteration of tessellations. Here, each cell of an initial tessellation is further subdivided into smaller cells by so–called component tessellations.

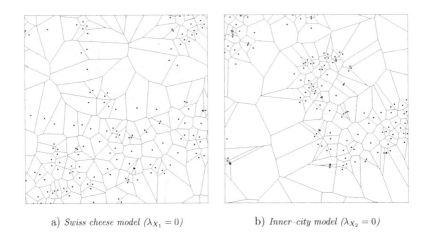

    a) Swiss cheese model ($\lambda_{X_1} = 0$)   *b) Inner–city model ($\lambda_{X_2} = 0$)*

Figure 2.14: Special cases of modulated Poisson–Voronoi tessellations with realisations of corresponding Boolean models in red

Let $\tau_{X_0} = \{\Xi_1, \Xi_2, ...\}$ be a stationary random tessellation in $\mathbb{R}^2$ and let $\{\tau_{X_n}\}_{n \geq 1}$ be a sequence of stationary tessellations $\tau_{X_n} = \{\Xi_{n_1}, \Xi_{n_2}, ...\}$ in $\mathbb{R}^2$ that are exchangeable that means

$$(\tau_{X_1}, ..., \tau_{X_n}) \overset{d}{=} (\tau_{X_{\pi(1)}}, ..., \tau_{X_{\pi(n)}})$$

for every n and every permutation $\pi : \{1, ..., n\} \rightarrow \{1, ..., n\}$. Moreover, let τ_{X_0} and $\{\tau_{X_n}\}_{n \geq 1}$ be independent. Then, the tessellation τ_X given by

$$\tau_X = \{\Xi_{n_\nu} \cap \Xi_n : \text{int } \Xi_{n_\nu} \cap \text{int } \Xi_n \neq \emptyset, \quad n, \nu = 1, 2, ...\}$$

is called an *iterated tessellation* in $\mathbb{R}^2$ with *initial tessellation* τ_{X_0} and *component tessellations* $\{\tau_{X_n}\}_{n \geq 1}$. More specifically, if the component tessellations $\{\tau_{X_n}\}_{n \geq 1}$ satisfy the condition that $\tau_{X_1} = \tau_{X_2} = ...$ we call the stationary iterated tessellation τ_X a *superposition*, whereas if the sequence $\{\tau_{X_n}\}_{n \geq 1}$ of component tessellations consists of independent and identically distributed stationary tessellations independent of τ_{X_0}, we call τ_X a *nesting*.

In Figure 2.15 some realisations of different types of one–fold superpositions are shown. Note that there is no hierarchy in the two tessellations involved in a superposition, this means that, for example, a PVT/PLT superposition has the same distribution as a PLT/PVT superposition with reversed parameters, and that a realisation of a superposition of a PLT with parameter γ_1 by another PLT with parameter γ_2 is identical to a realisation of a PLT with parameter $\gamma_1 + \gamma_2$.

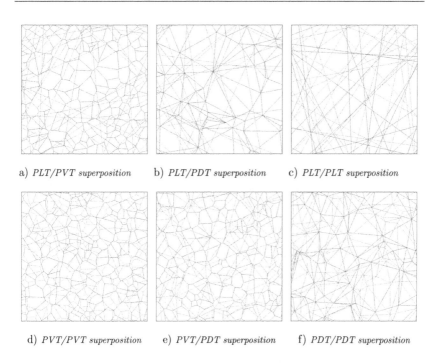

   a) *PLT/PVT superposition*   b) *PLT/PDT superposition*   c) *PLT/PLT superposition*

   d) *PVT/PVT superposition*   e) *PVT/PDT superposition*   f) *PDT/PDT superposition*

Figure 2.15: Realisations of one–fold superpositions using PVT, PDT and PLT as basic models shown in red

Figure 2.16 displays the nine possible models for a one–fold nesting that arise if the three basic models PVT, PDT, and PLT are possible choices for the initial tessellation as well as for the component tessellation of an iterated tessellation. Note that it is also possible to further extend the model of iterated tessellations. A natural extension of one–fold iterated tessellations is given by a k–fold iterated tessellation, where $k \geq 2$. This means that the cells of a $(k-1)$–fold iterated tessellation are further tessellated (cmp. Figure 2.17 for the cases of a two–fold superposition and a two–fold nesting). In this thesis we will only focus on one–fold iterated tessellations. Another extension possibility is to regard a nesting, where the component tessellations do not have to be identically distributed (but are still independent of each other). Such a model is called a *multitype nesting* of tessellations (cmp. [40], [92]). An important variation of a nesting for tessellations is given by the introduction of a Bernoulli probability p_B that represents the probability for an initial cell Ξ_n to be further iterated by a component tessellation. Figure 2.18 shows some examples of realisations of such a *Bernoulli thinning* with parameter $p_B = 0.75$ for different iterated tessellation models.

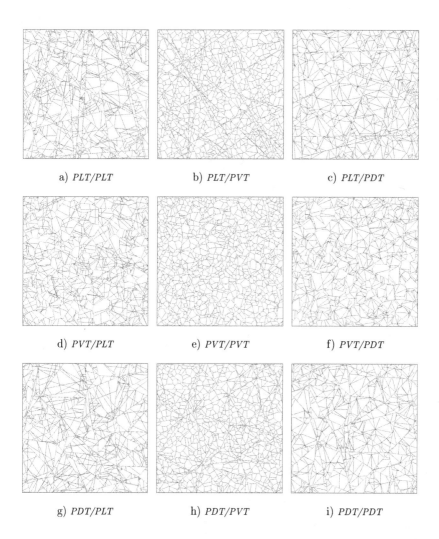

Figure 2.16: Realisations of one–fold nestings using PVT, PDT and PLT as basic models shown in red

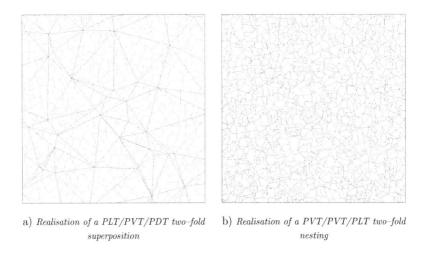

a) *Realisation of a PLT/PVT/PDT two–fold*   b) *Realisation of a PVT/PVT/PLT two–fold*
            superposition   *nesting*

Figure 2.17: Examples of two–fold iterated tessellations

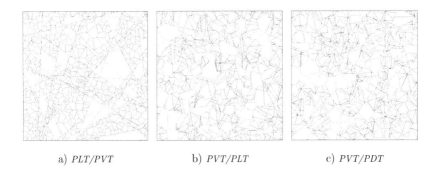

a) *PLT/PVT*   b) *PVT/PLT*   c) *PVT/PDT*

Figure 2.18: Realisations of one–fold nested tessellations with Bernoulli thinning ($p_B = 0.75$)

Chapter 3

Simulation Algorithms for the Typical Cell of Random Tessellations

A key component for an analysis of random tessellation models is the typical cell introduced in Section 2.4.2. Briefly, a typical cell Ξ_τ^* of a random stationary tessellation τ_X is a random polygon that is distributed according to the Palm mark distribution $P_{\tau_X}^*$ of the corresponding tessellation τ_X. More concrete, a realisation of the typical cell is obtained by drawing uniformly from the pool of all cells available for the specific tessellation model.

It is important to note that as long as the tessellation models considered are ergodic, which is the case for the examples we will have a look at, characteristics of the typical cell can be used to obtain inference about characteristics that are obtained from spatial averaging in large sampling windows like the mean perimeter of the cells. Hence, we are able to get information about spatial averages without having to simulate in large sampling windows and without having to face problems connected with such methods like occurring edge effects, computer memory problems and difficulties in parallelisation of computations. Another important advantage of a typical cell approach is that, since independent and identically distributed samples are obtained, we are able to apply well-known mathematical facts like the Central Limit Theorem (Theorem A.1) in order to obtain knowledge at least about the asymptotic behavior of certain estimators and statistics of interest.

All algorithms for simulations of the typical cell for random tessellation models introduced in this chapter are more or less based on the simulation of Poisson point processes. Therefore, in Section 3.1 techniques for the simulation of stationary as well as instationary Poisson point processes are discussed. Slivnyak's theorem is introduced that allows the construction of efficient algorithms for the simulation of the typical cell for random tessellations that are based on Poisson point processes. Finally, a simulation algorithm for the typical cell of a Poisson–Voronoi tessellation is discussed.

In Section 3.2 a simulation algorithm for the typical cell of a Cox–Voronoi tessellation defined in Section 2.4.6 is developed. Some results of Monte–Carlo simulations based on this algorithm are provided. Characteristics of interest for such Monte–Carlo simulations are for example the area, the perimeter or the number of vertices of the typical cell, in all three cases with respect to distributions and resulting moments.

In Section 3.3 it is described how to simulate the typical cell of a modulated Poisson–Voronoi tessellation defined in Section 2.4.7. Monte–Carlo simulations are performed in order to obtain inference about cell characteristics similar as in Section 3.2. Note that the simulation algorithms introduced in Sections 3.2 and 3.3 will be applied in Chapter 4 to derive estimations of cost functionals for models based on the random tessellations involved.

3.1   General Aspects

In this section algorithms for the simulation of Poisson point processes and for the simulation of the typical cell for a Poisson–Voronoi tessellation are summarized.

3.1.1   Simulation of Poisson Point Processes

With respect to the simulation of stationary Poisson point processes in $\mathbb{R}^2$ we can distinguish between two different techniques. If the sampling window is known beforehand a simulation algorithm based on the definition of the planar Poisson point process given in Section 2.2.4 can be used. In the case that the sampling window is undefined yet or that we do not want to restrict the simulation to a fixed sampling window a method called radial simulation is advisable. Both types of simulation are described in the following.

Simulation in a fixed and bounded sampling window
If the simulation is to be performed in a known fixed and bounded sampling window, the definition of a stationary Poisson point process introduced in Section 2.2.4 can directly be used for the construction of a simulation algorithm. More precisely, given an arbitrary but fixed and bounded sampling window $B \in \mathcal{B}_0(\mathbb{R}^2)$ we can realise a stationary Poisson point process X with intensity λ_X as follows. First, a Poisson distributed real–valued random variable $X(B)$ with $\mathbb{E}X(B) = \lambda_X \nu_2(B)$ is realised representing the number of points of X located in B. Afterwards for $i = 1, ..., X(B)$

the location of each point $X_i \in X$ is determined independently of each other and according to a uniform distribution on B. This means that if B is a rectangle of the form $[a_1, b_1] \times [a_2, b_2]$, where $a_1, b_1, a_2, b_2 \in \mathbb{R}$ and $a_1 < b_1$ and $a_2 < b_2$, the location of X_i is given as (X_{i_1}, X_{i_2}), where $X_{i_1} \sim U(a_1, b_1)$ and $X_{i_2} \sim U(a_2, b_2)$ are both uniformly distributed. For other shapes of B a rectangle B' can be regarded with $B \subset B'$ and a technique of rejection sampling can be applied, i.e., point proposals $X_i^* \in B'$ are generated but only kept as a point $X_i \in X$ if $X_i^* \in B$, otherwise the proposal is discarded and a new proposal for the i-th point is generated. By performing the simulation in this way it is assured that the points $X_1, ..., X_{X(B)}$ are all conditionally uniformly distributed on B since $P(X_i \in A | X_i \in B) = \nu_2(A)/\nu_2(B)$ for all subsets A of B and $i = 1, ..., X(B)$.

Radial simulation of stationary Poisson point processes

An alternative approach to the simulation of stationary Poisson point processes is given by radial simulation, where radial in this context means that the simulated points have an increasing distance to the origin. A more general description and mathematical details for the radial generation of Poisson point processes is provided in [86]. Recall that a point $x = (x_1, x_2) \in \mathbb{R}^2$ can be represented in polar coordinates as $x = (r, z)$, where $x_1 = r \cos z$ and $x_2 = r \sin z$. Consider a sequence of random variables $\{R_i\}_{i \geq 1}$ with $R_0 < R_1 < ...$ such that $\{R_i\}$ is a (linear) stationary Poisson point process with parameter γ. Furthermore, consider another sequence $\{Z_i\}_{i \geq 1}$ of independent and $U((0, 2\pi])$– distributed random variables, independent of $\{R_i\}$. Then the sequence $\{((R_i/\pi)^{1/2}, Z_i)\}$ is a (two–dimensional) stationary Poisson point process with parameter γ.

Hence, we can generate a stationary Poisson point process radially by simulating independent random variables $U_j \sim U(0, 1)$ and $V_i \sim U(0, 2\pi)$ and by putting

$$R_i = -\frac{1}{\gamma} \sum_{j=0}^{i} \log U_j \tag{3.1}$$

and

$$Z_i = V_i. \tag{3.2}$$

Simulation of instationary Poisson point processes

With respect to the simulation of instationary Poisson point processes a thinning method can be applied. Let X be an instationary Poisson point process with intensity measure Λ_X which has a representation of the form given in (2.10). Furthermore, for a bounded sampling window $B \in \mathcal{B}_0(\mathbb{R}^2)$ let $\lambda_{\max} = \max_{x \in B}\{\lambda_X(x)\}$ where we assume that $0 < \lambda_{max} < \infty$. Then we can generate realisations of X in the sampling window B as follows.

As it was described above, we generate a realisation of a stationary Poisson point process X' with intensity $\lambda_{\max}$ in B. Then, for each point x_i' of the realisation we perform a Bernoulli thinning procedure with parameter $p_i = \frac{\lambda_X(x_i)}{\lambda_{\max}}$. This means that x_i' is kept with probability p_i and discarded with probability $1 - p_i$. This Bernoulli thinning is done for each point independently. The point pattern that results from the points that are not discarded is then a realisation of the instationary Poisson point process X with intensity measure Λ_X.

3.1.2   Slivnyak's Theorem

For a stationary Poisson point process X it is possible to derive a relationship between the distribution P_X of the Poisson point process and its Palm distribution P_X^*. This relationship is given by Slivnyak's theorem ([98]).

Theorem 3.1 *(Slivnyak's theorem) Let X be a stationary Poisson point process with distribution P_X and Palm distribution P_X^*. Then the following holds*

$$P_X^*(X \in A) = P_X(X + \delta_o \in A), \ A \in \mathcal{N}, \tag{3.3}$$

where δ_o denotes the degenerate point process that consists only of a deterministic point in o.

A proof of Slivnyak's theorem can be found for example in [94], pp. 87f. Note that while in Chapter 2 the point process X has been defined on a general probability space $(\Omega, \mathcal{A}, \mathbb{P})$, we are now regarding the point process on the so-called canonical probability spaces $(N, \mathcal{N}, P_X)$ and $(N, \mathcal{N}, P_X^*)$, respectively.

Slivnyak's theorem provides a very useful representation for the Palm distribution of stationary Poisson point processes. Hence, we are able to simulate Palm distributions of stationary Poisson point process by simulating unconditional distributions of stationary Poisson point processes and by afterwards adding a point at the origin. A direct application for Slivnyak's theorem is the simulation of the typical cell for Poisson–Voronoi tessellations that is explained in the following.

3.1.3   Simulation of the Typical Cell for Poisson–Voronoi Tessellations

The simulation algorithm of the typical cell Ξ_τ^* for a Poisson–Voronoi tessellation τ_X that is introduced in the following is based on the algorithm explained in [86]. We will later on reuse some ideas of this algorithm for the simulation of the typical cell for other Voronoi–type tessellations. The algorithm uses the fact that due to the one–to–one correspondence between the Poisson–Voronoi cells and their nuclei it is useful for the simulation of the typical Poisson–Voronoi cell to simulate a point process according to the Palm distribution P_X^* of the generating Poisson point process X. In other words, by using Slivnyak's theorem introduced in Section 3.1.2, it holds that the typical cell of a Poisson–Voronoi tessellation τ_X induced by a Poisson point process X has the same distribution as the Voronoi cell with nucleus at o that is induced by the point process $X + \delta_o$. Therefore, we have to simulate the point process $X + \delta_o$ in order to obtain a Voronoi cell with nucleus at o that is equal (with respect to its distribution) to the typical cell of a Poisson–Voronoi tessellation generated by X. In order to derive such a simulation, we are able to use techniques of radial simulation for Poisson point processes introduced in Section 3.1.1. We first add a point X_0 at the origin. Then, a Poisson point process is radially simulated by generating random points $X_1, X_2, \ldots$ with an increasing distance to the origin according to the description given in Section 3.1.1.

Construction of an initial cell
An important problem in the efficient simulation of the typical cell Ξ_τ^* for Poisson–Voronoi tessellations is the construction of an initial cell. In other words, if we regard the bisectors of (X_0, X_i) for $i = 1, \ldots, n$ we are looking for the smallest n that fulfills the condition that X_0 is completely surrounded by a convex polygon formed by these bisectors. A procedure for the construction of the initial cell in the Poisson–Voronoi case is visualized in Figure 3.1. The lines $\overline{X_1 X_0}$ and $\overline{X_2 X_0}$ form with probability one a cone S_2 with respect to the opposite side of X_0. If the nearest point X_3 lies inside of this cone the algorithm stops and an initial cell can be constructed using the bisectors (X_0, X_1), (X_0, X_2) and (X_0, X_3). Otherwise the cone S_3 is taken as the maximal cone formed by two of the three lines $\overline{X_1 X_0}$, $\overline{X_2 X_0}$ and $\overline{X_3 X_0}$ on the opposite side of X_0. Afterwards the point X_4 is taken into account with respect to S_3 (Figure 3.1a). This procedure is repeated until $X_{i+1} \in S_i$. With probability one this algorithm stops after a finite number of steps (cmp. [113]) and an initial cell can be constructed by using the corresponding bisectors (Figure 3.1b). Note that there are other stopping criterions thinkable for Voronoi tessellations in general which might prove to be more efficient for specific situations, especially with regard to runtime optimization.

Stopping criterion and construction of the typical cell
After the creation of an initial cell a stopping criterion for the simulation of the typical cell Ξ_τ^* of a Poisson–Voronoi tessellation τ_X can be provided ([86]). If $d_{\max}$ denotes the maximal distance of the vertices for the initial cell to the origin (that is identical to X_0 in this case) then only points that are located inside the disc with radius $2d_{\max}$ centered at the origin can influence the shape of the typical cell. Hence, it is sufficient to simulate points $X_i \in X$ until $|X_i| > 2d_{\max}$. Note that each time a new point X_i is simulated it is possible that it cuts the initial cell, thereby modifying it and that thus also the maximal distance is reduced. The final result after fulfilling the stopping

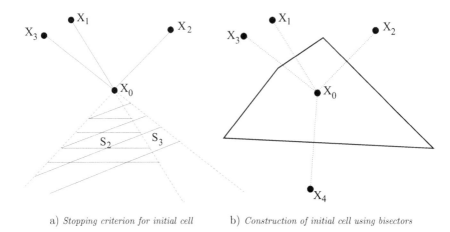

a) *Stopping criterion for initial cell*   b) *Construction of initial cell using bisectors*

Figure 3.1: Stopping criterion for an initial cell of the typical cell and its construction

criterion is a realisation ξ_τ^* of the typical cell Ξ_τ^* for a Poisson–Voronoi tessellation τ_X.

3.2 Cox–Voronoi Tessellations Based on Poisson Line Processes

In Section 2.4.6 a Cox–Voronoi tessellation τ_{X_c} has been introduced that is induced by a Cox point process X_c with random driving measure Λ_{X_c} given in (2.31). In this section we present an algorithm for the simulation of the typical cell of such a Cox–Voronoi tessellation τ_{X_c}. The algorithm will be used in Section 3.2.3 in order to derive estimations about characteristics of the typical cell by Monte–Carlo simulation. In Section 4.2 the algorithm will prove to be useful for the analysis of cost functionals in two–level hierarchical models based on two Cox point processes. The material presented in Section 3.2 is based partially on results obtained in [31].

3.2.1 Representation of the Typical Cell

As it is the case for the cells of a Poisson–Voronoi tessellation, we have a one–to–one correspondence between the cells of the Cox–Voronoi tessellation τ_{X_c} induced by the

Cox point process X_c and the generating Cox point process X_c itself that represents the nuclei of the cells. Therefore, in order to describe the distribution of the typical cell Ξ_τ for the Cox–Voronoi tessellation τ_{X_c}, it is useful to describe the Palm distribution $P^*_{X_c}$ of X_c which is done in the following.

Let X_c be a (stationary) Cox point process with random driving measure Λ_{X_c} given in (2.31) and with intensity $\lambda_c = \lambda_\ell \gamma$ that is induced by (linear) Poisson point processes with (linear) intensity λ_ℓ on the lines of a Poisson line process X_ℓ with intensity γ. Furthermore, let $\ell_{(o, Z_0)}$ be a line through the origin with random orientation angle Z_0 which is independent of X_c and uniformly distributed on $[0, 2\pi)$. Given Z_0, consider X' as a (linear) stationary Poisson point process on $\ell_{(o, Z_0)}$ with (linear) intensity λ, again independent of X_c. Then, the Palm distribution $P^*_{X_c}$ of X_c has the form

$$P^*_{X_c}(X_c \in A) = P_{X_c}(X_c + X' + \delta_o \in A), \ A \in \mathcal{N}, \qquad (3.4)$$

where δ_o is the degenerate point process that consists solely of the (deterministic) point o (cmp. also [93]). The palm representation given in (3.4) is based on the independent (and isotropic) placement of the lines of the underlying Poisson line process with respect to each other as well as on the subsequent independent placement of the points of X_c on these lines.

Hence, in order to simulate the typical cell Ξ^*_τ of the Voronoi tessellation τ_{X_c} induced by X_c we have to simulate the point process $X_c + X' + \delta_o$. Note that, analogously to (3.3), the processes X_c and $X_c + X'$ in (3.4) are regarded with respect to the canonical probability spaces $(\Omega, \mathcal{N}, P^*_{X_c})$ and $(\Omega, \mathcal{N}, P_{X_c})$, respectively. The Voronoi cell with o as its nuclei is then, with respect to its distribution, identical to the typical cell Ξ^*_τ of the Cox–Voronoi tessellation τ_{X_c} induced by X_c.

3.2.2   Simulation Algorithm

Using the representation of the typical cell Ξ^*_τ given in Section 3.2.1, we start our algorithm (Fig. 3.2), by adding a point X_0 at the origin o and by the simulation of an initial line $\ell_0 = \ell_{(o, Z_0)}$ passing through $X_0 = o$ with a uniform orientation on $[0, 2\pi)$. With respect to $X_0 = o$ on ℓ_0, the nearest neighbour points X_1 and X_2 in each direction of ℓ_0 then have Euclidean distances D_1 and D_2 from X_0, where D_1 and D_2 are independent and $Exp(\lambda)$–distributed due to the properties of the one–dimensional Poisson point process (Figure 3.2a).

Construction of the initial cell

In order to construct an initial cell we then simulate a second line ℓ_1. Recall that for the purpose of simulating a Poisson line process X_ℓ radially, i.e., with increasing distance to the origin, it suffices to simulate independent random variables $T_j \sim Exp(2\gamma)$ and

$Z_i \sim U[0, 2\pi]$ for each $i \in \{1, \ldots, k\}$ and for some $k \geq 1$ (cmp. Sections 2.4.3 and 3.1.1). Then, k simulated lines can be obtained from the pairs (R_i, Z_i), where $R_i = \sum_{j=1}^{i} T_j$. Therefore, a uniformly oriented second line $\ell_1 = \ell_{(R_1, Z_1)}$ is simulated, where $R_1 \sim Exp(2\gamma)$, and the point of intersection $P_{(\ell_0, \ell_1)}$ between ℓ_0 and ℓ_1 is computed. Then, the nearest neighbour points of $P_{(\ell_0, \ell_1)}$, say X_3 and X_4, are simulated on ℓ_1 using the memoryless property of the one–dimensional Poisson point process on ℓ_1. This means that the distances of the nearest neighbour points in each direction of ℓ_1 from the point of intersection $P_{(\ell_0, \ell_1)}$ are again $Exp(\lambda)$–distributed (Figure 3.2b). The four points X_1, X_2, X_3, and X_4, together with the origin X_0, can now be used in order to construct a first initial cell with nucleus at o by constructing the Voronoi cell of X_0 with respect to the set $\{X_1, X_2, X_3, X_4\}$ and their corresponding bisectors to X_0 (Figure 3.2c).

Stopping criterion and construction of the typical cell
By using the general construction principle of Voronoi tessellations and similarly as for the case of the simulation for the typical cell of the Poisson–Voronoi tessellation described in Section 3.1.3, the initial cell for the typical cell of a Cox–Voronoi tessellation τ_{X_c} induced by a Cox point process X_c provides an upper bound for the maximum distance from X_0 to all those lines of X_ℓ that can influence the shape of the Voronoi cell with X_0 as its nucleus. This maximum distance equals two times the maximum distance of all vertices for the initial cell from X_0 (Figure 3.2c). Note that it is not necessary to simulate further points located on ℓ_o with respect to the simulation of the typical cell, since these further points on ℓ_o are unable to influence the typical cell. This is due to the fact that all bisectors of points on ℓ_o with respect to X_0 are parallel and hence have no point of intersection with each other. For ℓ_1 this is not true, meaning that further points have to be simulated with an exponentially distributed distance to the adjacent point on ℓ_1 until the stopping criterion is met, i.e. until the distance of the simulated points becomes larger than two times the maximum distance of all the vertices of the initial cell from X_0. By simulating further lines $\ell_{i+1} = \ell_{(R_{i+1}, Z_{i+1})}, i \geq 1$ with $R_{i-1} < R_i$ and $R_i - R_{i-1} \sim Exp(2\gamma)$, and by simulating points of X_c on these lines until the stopping criterion is met, it is possible to generate a cell whose distribution coincides with the distribution of the typical cell Ξ_τ^* of X_c (Figure 3.2d). With respect to an efficient implementation it is advisable to adjust the new maximum distance after having simulated a new line with simulated points on it and after having constructed the corresponding bisectors with regard to X_0. More precisely, if the considered cell is split by a bisector of one of the newly simulated points, the regarded maximum distance can possibly be reduced. The whole procedure of reducing the maximum distance is carried out until the distance of the next simulated line from X_0 is bigger than the maximum distance, which is equal to two times the maximum distance from all vertices of the regarded cell to X_0.

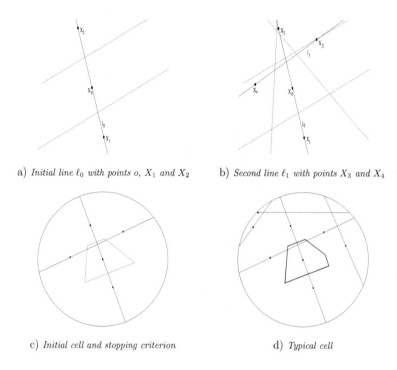

a) *Initial line ℓ_0 with points o, X_1 and X_2*   b) *Second line ℓ_1 with points X_3 and X_4*

c) *Initial cell and stopping criterion*   d) *Typical cell*

Figure 3.2: Simulation algorithm for the typical Cox–Voronoi cell Ξ_τ^* induced by X_c

3.2.3 Results of Monte–Carlo Simulations

We now present some results of Monte–Carlo simulations that have been obtained by an implementation of the algorithm developed in Section 3.2.2. These results were obtained in cooperation with M. Rösch and are also partially documented in his diploma thesis ([88]). Of particular interest for the typical cell Ξ_τ^* of a Cox–Voronoi tessellation τ_X that is induced by a Cox point process X_c are distributional properties as well as first–order and second–order moments of cell characteristics such as area, perimeter, and number of vertices. Additionally, we will examine differences in the behavior of corresponding characteristics for the typical cell of τ_{X_c} compared to the typical cell of a Poisson–Voronoi tessellation with the same intensity, i.e. in particular with the same mean area. Due to the scaling invariance property of τ_{X_c} that has been explained in Section 2.4.6 we mostly reduce ourselves to an examination of results for the one–dimensional parameter $\kappa = \gamma/\lambda$ instead of the vector (γ, λ).

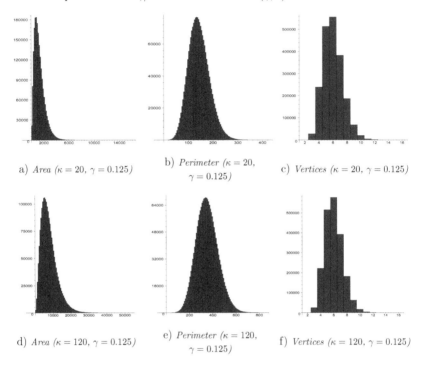

a) *Area* ($\kappa = 20$, $\gamma = 0.125$)

b) *Perimeter* ($\kappa = 20$, $\gamma = 0.125$)

c) *Vertices* ($\kappa = 20$, $\gamma = 0.125$)

d) *Area* ($\kappa = 120$, $\gamma = 0.125$)

e) *Perimeter* ($\kappa = 120$, $\gamma = 0.125$)

f) *Vertices* ($\kappa = 120$, $\gamma = 0.125$)

Figure 3.3: Histograms for characteristics of the typical Cox–Voronoi cell

Distributional properties
For the Monte–Carlo simulations leading to the results shown in the following we used $n = 2{,}000{,}000$ iterations. Although this seems to be quite a large number, for some specific evaluations, especially with respect to a comparison to characteristics for the typical cell of a Poisson–Voronoi tessellation, such a large number of iterations proved to be necessary in order to ensure sufficient accuracy. In Figure 3.3 histograms for the area $\nu_2(\Xi_\tau^*)$, the number of vertices $\eta(\Xi_\tau^*)$, and the perimeter $\nu_1(\partial\Xi_\tau^*)$ of the typical Cox–Voronoi cell Ξ_τ^* are displayed, where $\gamma = 0.125$ and $\kappa = 20$ or $\kappa = 120$, respectively. At first sight, the area in both cases seems to be similar to a γ–distribution (see

Appendix A.4), whereas the histograms for the perimeter of the typical cell look like a histogram of a normal distribution. The histograms of the number of vertices seem to have a similar shape compared to the ones for the area, but of course in a discrete setting. Note furthermore that the modes for histograms of the vertices coincide with the expected value $\mathbb{E}\eta(\Xi_\tau^*) = 6$ of the theoretical distributions. For other choices of the parameter κ, the histograms look very similar. Recall that due to the scaling invariance property, the histograms have identical shapes as long as κ remains fixed. Only the scales may change of course (except for the number of vertices, where even the scale stays the same).

First–order and second–order moments
In Table 3.1 results for functionals $f(\Xi_\tau^*)$ of the typical Cox–Voronoi cell Ξ_τ^* are shown, where the functional $f(\Xi_\tau^*)$ represents either the area $\nu_2(\Xi_\tau^*)$, the perimeter $\nu_1(\partial\Xi_\tau^*)$, or the number of vertices $\eta(\Xi_\tau^*)$. Apart from the expectations $\mathbb{E}f(\Xi_\tau^*)$, also the variances $\mathrm{Var}f(\Xi_\tau^*)$ as well as the coefficients of variation $\mathrm{cv}f(\Xi_\tau^*) = 100\sqrt{\mathrm{Var}f(\Xi_\tau^*)}/\mathbb{E}f(\Xi_\tau^*)$ (i.e., standard deviation times 100 divided by expectation) are displayed in Table 3.1. Results are shown for different values of γ and for fixed parameter $\kappa = \gamma/\lambda = 50$. Recall that for different values of γ and fixed κ, the moments $\mathbb{E}f(\Xi_\tau^*)$ and $\mathrm{Var}f(\Xi_\tau^*)$, respectively, are related to each other by scaling. For example, $\mathbb{E}\eta(\Xi_\tau^*)$ is independent of γ, whereas $\mathbb{E}\nu_1(\partial\Xi_\tau^*)$ and $\sqrt{\mathbb{E}\nu_2(\Xi_\tau^*)}$ behave linearly with respect to $1/\gamma$. The scaling properties mentioned are reflected by the simulated estimates given in Table 3.1. In particular, the coefficients of variation displayed in Table 3.1 are almost constant for different values of γ as long as κ is kept fixed. In Table 3.2, the dual case is considered, meaning that γ is fixed, while κ is variable. By the same scaling properties as mentioned above, the simulated estimates given in Table 3.2 can be used in order to compute estimates for $\mathbb{E}f(\Xi_\tau^*)$, $\mathrm{Var}f(\Xi_\tau^*)$, and $\mathrm{cv}f(\Xi_\tau^*)$ for any $\kappa \in \{10, 20, 30, 40, 50, 60, 90, 120\}$ and γ arbitrarily chosen. For example, for $\kappa = 40$ and $\gamma = 0.25$, we would get the estimates 6.002, 100.712, and 640.1525 for $\mathbb{E}\eta(\Xi_\tau^*)$, $\mathbb{E}\nu_1(\partial\Xi_\tau^*)$, and $\mathbb{E}\nu_2(\Xi_\tau^*)$, respectively. If we would like to consider some $\kappa \notin \{10, 20, 30, 40, 50, 60, 90, 120\}$, estimates could either be determined by interpolation or extrapolation from the data given in Table 3.2, or by simulation for the value of κ under consideration and for some fixed γ and, afterwards, for the desired values of γ, by using scaling properties. Moreover, looking at the estimates for $\mathbb{E}\eta(\Xi_\tau^*)$ given in Table 3.2, we see that these estimates are almost identical to 6 for any κ, which is in confirmation with the scaling invariance of $\mathbb{E}\eta(\Xi_\tau^*)$. The estimates for the variances $\mathrm{Var}\eta(\Xi_\tau^*)$ seem to slightly decrease for increasing scaling parameter κ. On the other hand, the estimates for expectations and variances of perimeter and area, respectively, are increasing for an increasing κ, whereas the estimates for the coefficients of variation decrease for an increasing κ.

Comparison to PVT
For the purpose of analysing characteristics of the typical cell Ξ_τ^* for a Cox–Voronoi tessellations τ_{X_c} it is very interesting to compare them to characteristics for the typical

Table 3.1: Estimates for first–order and second–order moments for $\kappa = 50$ and different values of γ

	γ	$\mathbb{E}f(\Xi_\tau^*)$	$\mathrm{Var}f(\Xi_\tau^*)$	$\mathrm{cv}f(\Xi_\tau^*)$
	0.125	6.000	1.892	22.925
	0.25	6.001	1.896	22.945
	0.4	5.998	1.896	22.957
$\eta(\Xi_\tau^*)$	0.5	5.999	1.897	22.959
	0.8	6.000	1.895	22.943
	1.0	6.001	1.896	22.945
	1.25	6.001	1.900	22.970
	1.5	6.001	1.900	22.970
	0.125	225.207	3912.919	27.776
	0.25	112.617	976.756	27.752
	0.4	70.370	382.203	27.782
$\nu_1(\partial\Xi_\tau^*)$	0.5	56.297	244.286	27.763
	0.8	35.205	95.521	27.762
	1.0	28.165	61.139	27.762
	1.25	22.540	39.168	27.766
	1.5	18.766	27.134	27.758
	0.125	3198.954	3747622.689	60.516
	0.25	799.828	233774.327	60.451
	0.4	312.300	35711.831	60.511
$\nu_2(\Xi_\tau^*)$	0.5	199.882	14625.775	60.504
	0.8	78.172	2234.666	60.472
	1.0	50.026	914.832	60.461
	1.25	32.040	375.760	60.501
	1.5	22.212	180.516	60.488

Table 3.2: Estimates for first–order and second–order moments for $\gamma = 0.125$ and different values of κ

	κ	$\mathbb{E}f(\Xi_\tau^*)$	$\mathrm{Var}f(\Xi_\tau^*)$	$\mathrm{cv}f(\Xi_\tau^*)$
	10	5.998	2.088	24.091
	20	6.001	1.981	23.454
	30	6.002	1.939	23.200
$\eta(\Xi_\tau^*)$	40	6.002	1.915	23.056
	50	5.999	1.892	22.929
	60	6.000	1.883	22.870
	90	5.999	1.863	22.752
	120	6.000	1.850	22.669
	10	100.500	1000.239	31.469
	20	142.271	1771.053	29.580
	30	174.355	2501.238	28.684
$\nu_1(\partial\Xi_\tau^*)$	40	201.424	3210.422	28.130
	50	225.207	3912.919	27.776
	60	246.843	4599.240	27.474
	90	302.432	6640.160	26.944
	120	349.528	8637.773	26.590
	10	639.216	197578.455	69.538
	20	1280.290	688685.388	64.819
	30	1920.488	1447118.677	62.638
	40	2560.610	2467092.919	61.340
$\nu_2(\Xi_\tau^*)$	50	3198.953	3747622.689	60.516
	60	3840.243	5272317.126	59.792
	90	5758.732	11386016.845	58.595
	120	7684.181	19751363.890	57.836

Table 3.3: Expected perimeters of $\Xi_{\tau'}^*$ and Ξ_τ^* provided that $\mathbb{E}\nu_2(\Xi_{\tau'}^*) = \mathbb{E}\nu_2(\Xi_\tau^*) = 100$

κ	γ	λ	λ_p	CVT	PVT
10	0.3162	0.03162	0.0100	39.731	40.000
20	0.4472	0.02237	0.0100	39.785	40.000
30	0.5477	0.01826	0.0100	39.793	40.000
40	0.6325	0.01581	0.0100	39.807	40.000
50	0.7071	0.01414	0.0100	39.832	40.000
60	0.77460	0.01291	0.0100	39.834	40.000
90	0.9487	0.01054	0.0100	39.848	40.000
120	1.095	0.00913	0.0100	39.879	40.000

cell $\Xi_{\tau'}^*$ for a suitable Poisson–Voronoi tessellation τ_X', where suitable means that the typical cell $\Xi_{\tau'}^*$ of the Poisson–Voronoi tessellation has the same mean area as the typical cell Ξ_τ^* of the Cox–Voronoi tessellation. Such a comparison is of interest since the Poisson–Voronoi tessellation can be considered as the limit (in the sense of weak convergence) of a sequence of Cox–Voronoi tessellations, having all the same mean area and monotonously increasing coefficient κ as $\kappa \to \infty$. Unfortunately, we are not able to observe characteristics for the typical cell of a Cox–Voronoi tessellation if κ becomes too large since this means that too many lines have to be simulated in order to obtain the tessellation. Therefore, we restrict the following analysis to $\kappa \leq 120$. Especially interesting is the comparison of the expected perimeter $\mathbb{E}\nu_1(\partial\Xi_\tau^*)$ of the typical cell Ξ_τ^* for a Cox–Voronoi tessellation with the expected perimeter $\mathbb{E}\nu_1(\partial\Xi_{\tau'}^*)$ of the typical cell $\Xi_{\tau'}^*$ of a Poisson–Voronoi tessellation with the same intensity. Same intensity here means that $\lambda_c = \lambda_p$, where $\lambda_c = \lambda\gamma$ is the intensity of the Cox line process and λ_p represents the intensity of the Poisson point process, respectively. Recall that for the typical cell $\Xi_{\tau'}^*$ of a Poisson–Voronoi tessellation with intensity λ_p it holds that (cmp. [78])

$$\mathbb{E}\nu_2(\Xi_{\tau'}^*) = \frac{1}{\lambda_p}, \quad \mathbb{E}\nu_1(\partial\Xi_{\tau'}^*) = \frac{4}{\sqrt{\lambda_p}}, \quad \mathbb{E}\eta(\Xi_{\tau'}^*) = 6.$$

In particular, $\mathbb{E}\eta(\Xi_{\tau'}^*) = \mathbb{E}\eta(\Xi_\tau^*)$ and, assuming that $\lambda_p = \lambda\gamma$, we have $\mathbb{E}\nu_2(\Xi_{\tau'}^*) = \mathbb{E}\nu_2(\Xi_\tau^*)$. Furthermore, we are able to compare the expected perimeter $\mathbb{E}\nu_1(\partial\Xi_\tau^*)$ to the estimate for $\mathbb{E}\nu_1(\partial\Xi_{\tau'}^*)$ obtained by the simulation algorithm described in Section 3.2.2. Some results of Monte–Carlo simulations are displayed in Tables 3.3 and 3.4, where the expected areas $\mathbb{E}\nu_2(\Xi_{\tau'}^*)$ and $\mathbb{E}\nu_2(\Xi_\tau^*)$ coincide, being equal to 100 and 625, respectively. Similar results are obtained for other values of $1/\lambda_p$, where we can observe the following qualitative behavior. Estimates for the expected perimeter of the typical Cox–Voronoi

Table 3.4: Expected perimeters of $\Xi_{\tau'}^*$ and Ξ_τ^* provided that $\mathbb{E}\nu_2(\Xi_{\tau'}^*) = \mathbb{E}\nu_2(\Xi_\tau^*) = 625$

κ	γ	λ	λ_p	CVT	PVT
10	0.1265	0.01265	0.00160	99.312	100.000
20	0.1789	0.00895	0.00160	99.407	100.000
30	0.2191	0.00730	0.00160	99.472	100.000
40	0.2530	0.00633	0.00160	99.518	100.000
50	0.2828	0.00566	0.00160	99.593	100.000
60	0.3098	0.00516	0.00160	99.598	100.000
90	0.3795	0.00422	0.00160	99.615	100.000
120	0.4382	0.00365	0.00160	99.699	100.000

cell increase with respect to an increasing scaling parameter κ but it seems that they are in any case smaller than the expected perimeter of the typical cell of a Poisson–Voronoi tessellation of the same intensity. It might be difficult to come up with an analytical proof of this fact but a possible explanation of this interesting behavior could be that the typical cell of the Cox–Voronoi tesellation is more regularly shaped than the typical cell of the Poisson–Voronoi tessellation, because two edges of the typical Cox–Voronoi cell can be parallel with some positive probability. In the case of a Poisson–Voronoi tessellation this probability equals zero.

3.3 Voronoi Tessellations Based on Modulated Poisson Point Processes

A third kind of the typical cell for Voronoi tessellations is the typical cell Ξ_τ^* of a Voronoi tessellation τ_X that is induced by a modulated Poisson point process X with random driving measure Λ_X introduced in Sections 2.3.2 and 2.4.7. Apart from being an interesting mathematical object to analyse, the typical cell of Voronoi tessellations based on such modulated Poisson point processes has, for example, a considerable importance in the modelling of telecommunication networks on a nationwide scale. Here, they can represent structures like typical serving zones of antennas in mobile scenarios or of Wired Center Stations, a specific type of telecommunication equipment in access networks. The results of Section 3.3 are partially based on results obtained in [28].

3.3.1   Representation of the Typical Cell

We can describe the typical cell Ξ_X^* of a modulated Poisson–Voronoi tessellation τ_X as follows. Let the modulated Poisson point process X have random driving measure Λ_X given in (2.23) and let Ψ be the corresponding Boolean model. Consider the Palm distribution Q^* of the stationary random measure Λ_X (cmp. p. 229 in [100]) and denote by X^* a Cox point process with random driving measure Λ_{X^*} having distribution Q^*, where

$$Q^*(\cdot) = \frac{\lambda_{X_1}}{\lambda_X}\mathbb{P}(\Lambda_X \in \cdot, o \in \Psi) + \frac{\lambda_{X_2}}{\lambda_X}\mathbb{P}(\Lambda_X \in \cdot, o \notin \Psi). \qquad (3.5)$$

The Palm distribution Q^* of Λ_{X^*} given in (3.5) has an alternative representation as (cmp. also [93])

$$Q^*(\cdot) = p_c\mathbb{P}(\Lambda_X \in \cdot \mid o \in \Psi) + (1 - p_c)\mathbb{P}(\Lambda_X \in \cdot \mid o \notin \Psi), \qquad (3.6)$$

where

$$p_c = \frac{p_\Psi \lambda_{X_1}}{\lambda_X} \qquad (3.7)$$

is the conditional coverage probability $p_c = P_X^*(o \in \Psi)$ of the origin o by the Boolean model Ψ with respect to the Palm probability measure P_X^*, which means conditional to the event that at the origin a point of X is located. Recall that p_Ψ was defined in (2.22) and that it represents the (unconditional) coverage probability of o by the Boolean model Ψ, whereas λ_X is the intensity of X given in Lemma 2.6. Note that (3.6) can be obtained by an application of Bayes' Rule (cmp. p. 21 in [19]) and that (3.5) follows from (3.6) by using the definition of conditional probabilities. Note furthermore that in (3.5) as well as in (3.6) we are using the general probability space $(\Omega, \mathcal{A}, \mathbb{P})$ due to the fact that we have two random objects (Λ_X and Ψ) referring to it.

With respect to its Palm probability measure P_X^*, the Cox point process X has then the same distribution as $\delta_o + X^*$ has with respect to the original probability measure, i.e.,

$$P_X^*(X \in \cdot) = \mathbb{P}(X^* + \delta_o \in \cdot). \qquad (3.8)$$

We can deduce the representation provided in (3.8) from the fact that, given a realisation η of the random driving measure of X, the points of X are placed independently of each other which allows a representation similar to Theorem 3.1. Thus, we obtain that the typical cell Ξ_τ^* of τ_X has the same distribution as the Voronoi cell with nucleus at o that is induced by the point process $X^* + \delta_o$.

3.3.2   Simulation Algorithm

By using (3.5)–(3.7) we are able to obtain a theoretical basis for the simulation of the typical cell Ξ_τ^* for a modulated Poisson–Voronoi tessellation τ_X induced by a modulated

Poisson point process X. It is indicated by (3.8) that the typical cell of τ_X is equivalent (with respect to its distribution) to the Voronoi cell with nucleus at o of the point process $X^* + \delta_o$. Therefore, a way to get a realisation of the typical cell of τ_X is to simulate the modulated Poisson point process $X^* = \{X_n^*\}_{n \geq 1}$ with a random driving measure that has distribution Q^* given in (3.5). Note that due to (3.5) in order to simulate X^* we have to simulate the Boolean model Ψ^*, conditional to the events that the origin is covered by Ψ or not. This means that Ψ^* is simulated conditional to the event that $o \in X$, thereby inducing corresponding probabilities for $o \in \Psi$ and $o \notin \Psi$ according to (3.7). The simulation of X^* and Ψ^* is performed radially, i.e., with increasing distance to o and in an alternating fashion between the points of X^* and the germs of Ψ^*. As an initial step a point X_0^* is placed at the origin (Figure 3.4a). Thereby, the degenerate point process δ_o consisting of a (deterministic) point in o is represented (cmp. 3.8). Afterwards, it is determined by a Bernoulli experiment with success probability p_c defined in (3.7) whether X_0^* is covered by Ψ or not. In case that $o \in \Psi$ the distance of Y_1^*, the germ of Ψ^* which is nearest to X_0, to the origin has to be less than or equal to r, the (fixed) radius of the germs of Ψ^*. Otherwise, i.e., in case that $o \notin \Psi$, the distance of Y_1^* to o has to be bigger than r. Therefore, we have to simulate the distance of the first germ Y_1^* to the origin conditional to $o \in \Psi$ or $o \notin \Psi$, respectively (Figure 3.4b). A practical way to perform such a conditional simulation is to generate a proposal distance R_1 of the first germ Y_1^* to the origin according to (3.1) with $\gamma = \beta$. The proposal distance is accepted or rejected based on the conditions $R_1 \leq r$ or $R_1 > r$, respectively. In case of a rejection another proposal distance R_1 of Y_1^* to o is generated and the procedure is repeated until a proposal distance is accepted. After the generation of the distance R_1 of Y_1^* to o an angle Z_1 is simulated according to (3.2). Then, further points $X_i = (R_i, Z_i)$ are simulated radially by using (3.1) and (3.2) with intensity $\gamma = \max\{\lambda_{X_1}, \lambda_{X_2}\}$. For each of these points it is checked whether it is covered by Ψ^* or not (Figure 3.4c). This check is performed by simulating further germs Y_j^* of Ψ^* until either the distance of a germ to X_i becomes smaller than or equal to r or if the distance of Y_j^* to o becomes greater than $|X_i| + r$, where $|X_i|$ is the distance of X_i to the origin. In the first case, clearly, X_i is covered by Ψ^*, in the second it is not. After we have checked whether X_i is covered by the conditional Boolean model Ψ^*, in one of the two cases a thinning procedure has to be performed (cmp. Section 3.1.1). So, if without loss of generality $\lambda_1 > \lambda_2$ and $X_i \notin \Psi^*$ then the probability of discarding X_i is given by $1\lambda_2/\lambda_1$. Altogether, this method simulates $X^* = \{X_n^*\}_{n \geq 1}$ by an alternating simulation of a stationary Poisson point process $X_{\max}$ with intensity $\gamma = \max\{\lambda_{X_1}, \lambda_{X_2}\}$ and a conditional Boolean model Ψ^* and by applying the thinning procedure described above (Figure 3.4d). Note that an unconditional simulation of a (stationary) modulated Poisson point process in the plane can be done in a similar way by an alternating radial simulation of $X_{\max}$ and the (unconditional) Boolean model Ψ.

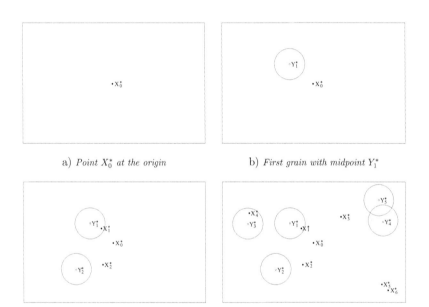

a) *Point X_0^* at the origin*   b) *First grain with midpoint Y_1^**

c) *For X_1 all necessary information about Ψ^* is given since $X_1 \in \Psi^*$. For X_2 more information* d) *Further alternating simulation of X^* and Ψ^* about Ψ^* is needed*

Figure 3.4: Algorithm, initial steps and alternating simulation of X^* and Ψ^*

Construction of an initial cell

With respect to the construction of an initial cell for the typical cell of a Voronoi tessellation that is induced by a modulated Poisson point process we are able to apply methods analogously to the case of a simulation algorithm for the typical cell of a Poisson–Voronoi tessellation. This means that as it has been explained in Section 3.1.3 given a nucleus X_0^* and two other points X_1^* and X_2^*, with probability one, we are able to construct a cone S_2 that is formed by the lines $\overline{X_1^* X_0^*}$ and $\overline{X_2^* X_0^*}$ with respect to the opposite side of X_0^* (Figure 3.1a). If X_3^*, the next point to be considered, is located in S_2 an initial cell can be constructed. Otherwise a new cone S_3 is taken as the maximal cone formed by two of the three lines $\overline{X_1^* X_0^*}$, $\overline{X_2^* X_0^*}$ and $\overline{X_3^* X_0^*}$ on the opposite side of X_0^*. Afterwards another point X_4^* is considered and the procedure is repeated until the cone S_i is finally hit by a point X_{i+1}^* which happens with probability one after a finite number of steps. Then an initial cell can be constructed using the bisectors (Figure 3.1b).

Stopping criterion

The stopping criterion for the simulation of the typical cell for a Voronoi tessellation that is induced by a modulated Poisson point process is analogous to the stopping criterion for the typical cell of Poisson–Voronoi tessellations given in Section 3.1.3. If $d_{\max}$ denotes the maximal distance of the vertices of the initial cell to the origin (that is identical to X_0) then the simulation of the points $X_i^* \in X^*$ has to be continued until the distance of X_i^* is bigger than $2d_{\max}$ (Figure 3.5). Note that $d_{\max}$ might be reduced during alterations of the cell (Figures 3.5b and 3.5c) and that therefore the stopping criterion should be adapted accordingly in order to ensure faster runtimes. The final result after fulfilling the stopping criterion is a realisation ξ_τ^* of the typical cell Ξ_τ^* for a modulated Poisson–Voronoi tessellation τ_X (Figure 3.5d).

Modifications for random radii

So far we simulated the typical cell Ξ_τ^* of Voronoi tessellations τ_X induced by a modulated Poisson point process X with respect to the assumption that the grains of the underlying Boolean model Ψ have a fixed radius. If instead the radius R of the circular grains of Ψ is random but bounded, for example, if $R \sim U[r - \delta, r + \delta]$ with $0 < \delta < r$, two modifications to the algorithm already introduced for a deterministic radius r have to be applied. It is important to note that, with respect to the simulation of the modulated Poisson point process X^*, in the case that the origin is covered by the conditional Boolean model Ψ^*, the grain generated by the first germ Y_1^* of Ψ^* with random radius R_1^* needs not necessarily cover the origin o. However, it is possible that another grain covers o. Therefore, after determining whether $X_0^* = o$ is covered by Ψ^*, the conditional radial simulation of the distances of the germs of Ψ^* to the origin together with the radii of the corresponding grains has to be performed in a way such that in the case $o \in \Psi^*$ at least one grain $Y_i^* + M_i^*$ covers X_0^*. On the other hand, if $o \notin \Psi^*$ one has to simulate grains $Y_i^* + M_i^*$ that do not cover X_0^* until the distance of

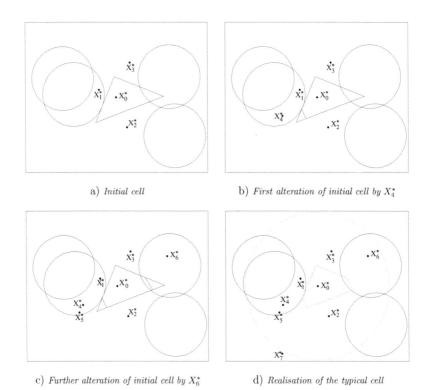

a) *Initial cell*

b) *First alteration of initial cell by X_4^**

c) *Further alteration of initial cell by X_6^**

d) *Realisation of the typical cell*

Figure 3.5: Alterations of initial cell and final realisation of the typical cell

their corresponding germs to X_0^* is larger than the maximal possible radius $r_{\max}$ (in the example of uniform distribution above $r_{\max} = r + \delta$). For the simulation algorithm this means that given $o \in \Psi^*$ or $o \notin \Psi^*$ a proposal sequence of germs $\{Y_i + M_i\}$ is radially generated for $i = 0, .., I_{\max}$, where $|Y_{I_{\max}}| > r_{\max}$. Afterwards it is checked whether this sequence fulfills the given condition or not. In the first case the sequence is accepted and the simulation of the grains is continued radially with the grain $Y_{I_{\max}+1} + M_{I_{\max}+1}$, otherwise a new proposal sequence is radially generated by starting at the origin again. This procedure is repeated until a sequence is found that can be accepted. An analogous modification has to be performed with respect to the necessary amount of grains that have to be simulated in order to know if a point X_i is covered by Ψ^* or not. In the case of a deterministic radius r of the grains it suffices to simulate until the distance of the germs of Ψ^* to the origin is bigger than $|X_i| + r$. Now for random radii, the necessary distance to the origin has to be bigger than $|X_i| + r_{\max}$, where again $r_{\max}$ is the maximal possible radius ($r + \delta$ in the example).

3.3.3 Results of Monte–Carlo Simulations

With regard to numerical evaluations of characteristics for the typical cell Ξ_τ^* of the Voronoi tessellation τ_X induced by a modulated Poisson point process X considered in this section a first statement that can be made is that due to the relatively large number of parameters needed, a complete analysis is almost impossible to achieve. Therefore we only concentrate on some specific scenarios to show some of the interesting effects that appear. The results of Section 3.3.3 have been obtained in cooperation with K. Posch and are also partially documented in her diploma thesis ([84]).

Transition to Swiss cheese model
The scenario we want to consider as a first example consists of a transition to a Swiss cheese model meaning that $\lambda_{X_1} \to 0$ while the intensity λ_X given in Lemma 2.6 is kept fixed. In particular we consider few large grains, where the (unconditional) coverage probability of the Boolean model Ψ is given by $p_\Psi = 0.6$ and where the intensity of the germs of Ψ is given by $\beta = 0.2$. Note that by using this parameter values and by applying (2.22) we then obtain a fixed radius $r = 1.20761$. We assume a fixed intensity $\lambda_X = 12$ such that the mean area of the typical cell remains fixed as $\mathbb{E}\nu_2(\Xi_\tau^*) = \lambda_X^{-1} = 0.8333$ (cmp. Lemma 2.8). We now let the parameter λ_{X_1} tend to 0 and regard the behavior of the distribution of the perimeter for the typical cell. Histograms of some sample cases for the choice of λ_{X_1} and λ_{X_2} are displayed in Figure 3.6, where the bars of the histograms have a width of 0.05 and with respect to each pair $(\lambda_{X_1}, \lambda_{X_2})$ a sample size of $n = 2{,}000{,}000$ is regarded. For the case of a stationary Poisson–Voronoi tessellation (Figure 3.6a), where $\lambda_{X_1} = \lambda_{X_2}$, a symmetrical look of the histogram for the perimeter of the typical cell Ξ_τ^* can be observed. This

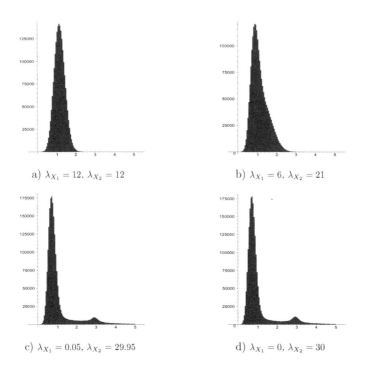

a) $\lambda_{X_1} = 12$, $\lambda_{X_2} = 12$ b) $\lambda_{X_1} = 6$, $\lambda_{X_2} = 21$

c) $\lambda_{X_1} = 0.05$, $\lambda_{X_2} = 29.95$ d) $\lambda_{X_1} = 0$, $\lambda_{X_2} = 30$

Figure 3.6: Histograms for the perimeter of the typical cell Ξ_τ^*

behavior is changing as λ_{X_1} tends to 0 resulting in a shape being skewed to the left. Another interesting effect that can be noticed is the existence of a second local maximum for the histogram, especially in the case of very small values for λ_{X_1}. This second local maximum is mainly due to cells that cover the grains of the corresponding Boolean model since inside of the Boolean model there are almost no points located if λ_{X_1} tends to be small.

Random radii of the grains
In a second example we regard a scenario where the radii are no longer fixed, but random. In particular, we consider fixed intensities $\lambda_{X_1} = 30$ and $\lambda_{X_2} = 9$. The random radius R of the grains of the Boolean model is given as $R \sim U(r - \delta, r + \delta)$, where r and δ are chosen such that $\mathbb{E}(R^2) = 4.5$ and $\delta < r$. Together with the choice of the intensity of the germs of the Boolean model as $\beta = 0.05$ we obtain by (2.22) the

Table 3.5: Mean values of characteristics for the typical cell Ξ_τ^* if R is random

r	δ	$\nu_2(\Xi_\tau^*)$	$\eta(\Xi_\tau^*)$	$\nu_1(\partial\Xi_\tau^*)$
1.84	1.83	0.05092	6.00049	0.87061
1.88	1.70	0.05091	5.99894	0.87066
1.92	1.56	0.05090	5.99863	0.87062
1.96	1.41	0.05092	6.00023	0.87093
2.00	1.22	0.05090	5.99915	0.87079
2.04	1.01	0.05090	5.99969	0.87082
2.08	0.72	0.05092	6.00001	0.87111
2.12	0.00	0.05090	5.99814	0.87110

unconditional coverage probability $p_\Psi = 0.50681$ and the intensity of the modulated Poisson point process $\lambda_X = 19.64298$. We now let r tend towards $\sqrt{4.5}$ under the condition that $r \leq \sqrt{4.5}$ which is equivalent to δ tending to 0 under the condition that $\delta \geq 0$. Results for mean characteristics obtained by simulations with $n = 5,000,000$ for each pair of parameters (r,δ) are displayed in Table 3.5. By looking at the estimated values for the mean area of the typical cell $\nu_2(\Xi_\tau^*)$ and the mean number of vertices $\eta(\Xi_\tau^*)$ we can deduce that the expected theoretical values of $\mathbb{E}(\nu_2(\Xi_\tau^*)) = 1/\lambda_X = 0.05091$ and $\mathbb{E}(\eta(\Xi_\tau^*)) = 6$ are quite well repeated, while for the mean perimeter of the typical cell $\nu_1(\partial\Xi_\tau^*)$ it seems to be the case that for less randomness (small δ) the mean perimeter seems to slightly increase.

Chapter 4

Estimation of Cost Functionals for Random Tessellation Models

In the following chapter we apply the results derived in Chapter 3 in order to efficiently estimate cost functionals connected to hierarchical models based on the random tessellation models regarded. More precisely, two–level hierarchical models are investigated, where both levels are based on the same type of tessellation, either the Cox–Voronoi tessellation induced by the Cox point process X_c introduced in Section 2.4.6 or the Voronoi tessellation induced by a modulated Poisson point process X with random driving measure Λ_X given in (2.23). For the hierarchical models inspected it is shown how to derive efficient estimators that are able to estimate specific cost functionals like the shortest path length or the Euclidean distance to the nearest element of higher order without having to simulate any lower–level elements. In order to derive these estimators Neveu's exchange formula for Palm distributions is applied that has been introduced in Section 2.2.7. In the first part of this chapter we recall some basic notions of graph theory. Algorithms for the computation of shortest paths and their corresponding path lengths will also be briefly discussed. Afterwards, in Section 4.2, we introduce the first two–level hierarchical model that is based on two Cox point processes in the sense of Section 2.4.6 that have a common underlying line process X_ℓ and that are independent of each other, given X_ℓ. With respect to this two–level hierarchical model efficient estimators for the mean shortest path length as well as for the mean subscriber line length are derived based on the simulation of the typical Voronoi cell Ξ_τ^* given in Section 3.2 that is induced by the upper–level Cox point process and on Neveu's exchange formula for Palm distributions introduced in Section 2.2.7. An analysis of the results for Monte–Carlo simulations of the model is provided that enables a calculation of the considered cost functionals for any given vector of parameters (γ, λ) without any further simulation necessary.

Another two–level hierarchical model that is regarded is based on two modulated Poisson point processes that have a common underlying Boolean model Ψ, but are independent of each other, given Ψ. Here, a cost functional of interest is the mean distance of a point belonging to the lower–level point process to its nearest neighbour that belongs to the point process of higher level. In order to derive an estimator for this cost functional the simulation of the typical cell Ξ_τ^* described in Section 3.3 for a Voronoi tessellation τ_X that is induced by a modulated Poisson point process as well as Neveu's exchange formula for Palm distributions are utilized. Results of estimations based on Monte–Carlo simulations are provided.

4.1   Graphs and Shortest Paths

In this section basic notions of graph theory are briefly recalled. For more detailed information the reader is, for example, referred to [23] and [42].

4.1.1   Definition of a Graph

First we want to define a graph for a given set of vertices. For this purpose we consider a non–empty set of vertices or nodes V (most of the times $V \subset \mathbb{R}^2$) and a non–empty set E of edges that connects exactly two such nodes (not necessarily different). Additionally, we consider a mapping $\alpha : E \to V \times V$ that assigns to an edge in E a pair of nodes in V. Then we call the triple $G = (V, E, \alpha)$ a (directed) *graph*.

Let $G = (V, E, \alpha)$ be a directed graph and consider a specific edge $e \in E$. If $\alpha(e) = (u, v)$ for some $u, v \in V$ the node u is called the *initial* node of e, while v is called *terminal* node of e. If there exists such an edge e, i.e., $\alpha(e) = (u, v)$, the node v is said to be a direct *successor* of u and u is said to be a direct *ancestor* of v.

A directed graph $G = (V, E, \alpha)$ is called *simple* if α is one–to–one, this means that there are no multiple edges between two nodes. Furthermore, G is called *complete* if $\alpha : E \to M_0$ is a surjective mapping, where $M_0 = \{(u, v);\ u \neq v;\ u, v \in V\}$. The graph G is called *finite* if V and E are finite.

The edge e having the property that $\alpha(e) = (u, u)$ for a $u \in V$ is called a *loop*. Thereby we can state that a simple graph does not contain any loops. A finite simple directed graph is called a *digraph*. In the following only digraphs of the form $G = (V, E, \alpha)$ are regarded, where $V = \{v_1, ..., v_m\}$.

An important type of graphs are weighted graphs, where a cost function is assigned to each edge of the graph. A mapping $c : E \to \mathbb{R}$ is called a *cost* or *weight function*. We call a digraph $G = (V, E, \alpha)$ together with a corresponding cost function c a *weighted digraph*. Often the cost function c can be written as a *cost matrix* $C = (c_{ij})$, where

$$c_{ij} = \begin{cases} 0 & \text{if} \quad i = j, \\ c(e) & \text{if} \quad \alpha(e) = (v_i, v_j), \\ \infty & \text{if} \quad \alpha^{-1}(v_i, v_j) = \emptyset. \end{cases}$$

Note that, for example if the digraph is not simple such a representation is not possible.

Now we turn to the notion of paths and in particular shortest paths that will play an important role in Section 4.2. Let $G = (V, E, \alpha, c)$ be a weighted digraph and consider a sequence $P = (e_1, e_2, ..., e_r)$. Suppose that the following conditions hold

1. $\alpha(e_1) = (u, x)$, $x \in V$,

2. $\alpha(e_r) = (y, v)$, $y \in V$,

3. $e_i \in E$, $i = 1, ..., r$,

4. The initial node of e_i is the terminal node of e_{i-1}, $i = 2, ..., r$.

Then P is called a *path* from u to v in G. A path that leads from a vertex $v \in V$ to the identical vertex v is called a *cycle*. Often it is not specifically the path $P = (e_1, e_2, ..., e_r)$ that is of interest but the *path length* $c(P)$ of P which is given as

$$c(P) = \sum_{i=1}^{r} c(e_i).$$

If a path P^* from u to v in G has minimal path length with respect to all possible paths from u to v in G we call P^* the *shortest* or *optimal path* from u to v in G. It is important to note that such a shortest path needs not always to exist. For example, it is possible that some paths might contain loops of negative length. In such cases it is easy to see that for any given constant $c \in \mathbb{R}$ there exists a path P such that $c(P) < c$. If two or more paths from u to v in G have the same minimal length with respect to all paths from u to v in G, usually one of them is chosen to be the shortest path.

Shortest paths and their lengths can be utilized to define the distance matrix as well as the path matrix of a graph. Let $G = (V, E, \alpha, c)$ be a weighted digraph without cycles of negative length. Denote by k_{ij} the shortest path length between two elements v_i and v_j of V. The matrix $D = \{d_{ij}\}$ with elements

$$d_{ij} = \begin{cases} 0 & \text{if} \qquad\qquad i = j, \\ k_{ij} & \text{if} \quad i \neq j \text{ and there exists a path between } v_i \text{ and } v_j, \\ \infty & \text{if} \qquad\quad \text{there is no path between } v_i \text{ and } v_j, \end{cases}$$

is called the *distance matrix* of G. The matrix $W = \{w_{ij}\}$ with elements

$$
w_{ij} = \begin{cases} i & \text{if} & i = j, \\ k & \text{if} & v_k \text{ is the direct ancestor of } v_j \text{ on the shortest path from } v_i \text{ to } v_j, \\ 0 & \text{else}, \end{cases}
$$

is called the *path matrix* of G. Note that if the path matrix is known it is possible to compute the shortest path between two vertices of the graph by a recursive determination of the edges involved.

4.1.2   Shortest Path Algorithms

Given a weighted digraph there are several algorithms known that can compute shortest paths and their corresponding lengths. In principle one can differ between single–source shortest path algorithms, where shortest paths are computed from a specific vertex to all other vertices and multi–source shortest path algorithms, where shortest paths are computed between all the vertices in the graph. In the following representants for both types of shortest path algorithms, Dijkstra's algorithm (single–source) and the Floyd–Warshall algorithm (multi–source), will be explained in detail. As denotations we will use V, the set of vertices or nodes; a the index of the initial vertex in V; v_a, the initial vertex; $m = \operatorname{card} V$, the number of vertices in V; $N(v_i)$, the set of successors of the vertex v_i, and c_{ij}, the associated costs of edge $e = (v_i, v_j)$. Dijkstra's algorithm is one of the most common algorithms for single–source shortest path algorithms. A small drawback is that this algorithm requires exclusively non–negative edge costs, hence it is only applicable for specific graphs. Effectively, Dijkstra's algorithm constructs a shortest path spanning tree by constructing a sequence of sets M_k with $M_k \subset V$ and sequences (with respect to the index k) of vectors $d_j^{(k)}$ and $w_j^{(k)}$ in $\mathbb{R}^m$ for $k, j = 1, ..., m$. In a first step M_1, $w^{(1)}$ and $d_j^{(1)}$ are initialised as

$$
\begin{aligned}
M_1 &= V \setminus \{v_a\}, \\
w_j^{(1)} &= \begin{cases} a & \text{if } (v_a v_j) \text{ exists}, \\ 0 & \text{else}, \end{cases} \\
d_j^{(1)} &= c_{aj}, \qquad 1 \leq j \leq m.
\end{aligned}
$$

In a second step for $k = 2, ..., m$ a vertex $v_i \in M_{k-1}$ is determined that fulfills the condition

$$
d_i^{(k-1)} = \min_{j:v_j \in M_{k-1}} d_j^{(k-1)}.
$$

The new set M_k is then given as $M_k = M_{k-1} \setminus \{v_i\}$, while for each $v_j \in M_{k-1}$ we put

$$
\begin{aligned}
w_j^{(k)} &= \begin{cases} i & \text{if} & d_j^{(k)} < d_j^{(k-1)}, \\ w_j^{(k-1)} & \text{else}, \end{cases} \\
d_j^{(k)} &= \min\{d_j^{(k-1)}, d_i^{(k-1)} + c_{ij}\}.
\end{aligned}
$$

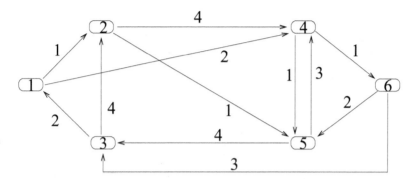

Figure 4.1: Sample graph for Dijkstra's algorithm

Under the necessary assumption that the weighted digraph has only non–negative edge costs the algorithm ends after at most m steps or as soon as $M_k = \emptyset$. In this case we obtain that

$$
\begin{aligned}
d_j^{(m)} &= \text{cost or length of the shortest path from } v_a \text{ to } v_j, \\
w_j^{(m)} &= \text{index in } V \text{ of the direct ancestor of } v_j \text{ along this path.}
\end{aligned}
$$

Therefore the costs of the shortest paths from the initial vertex v_a to all other vertices as well as the shortest paths themselves can be easily obtained from the two vectors $d^{(m)} = (d_1^{(m)}, ..., d_m^{(m)})$ and $w^{(m)} = (w_1^{(m)}, ..., w_m^{(m)})$. As a computational example regard the graph given in Figure 4.1, where v_4 is assumed to be the initial vertex. Applying Dikstra's algorithm we obtain the results displayed in Table 4.1 which, for example, tell us that the shortest path from v_4 to v_3 has a length of 4 and goes from v_4 to v_6 until it reaches v_3.

Floyd–Warshall algorithm

As an example for a multi–source shortest path algorithm we consider the Floyd–Warshall algorithm. Here, the shortest paths from each node to each other node are computed in a single algorithm. As an initial step two matrices $D_0 = \{d_{ij}^{(0)}\}$ and $W_0 = \{w_{ij}^{(0)}\}$ are constructed as

$$
\begin{aligned}
d_{ij}^{(0)} &= c_{ij}, \\
w_{ij}^{(0)} &= \begin{cases} i & \text{if} \quad i = j \text{ or } e = (v_i v_j) \text{ exists,} \\ 0 & \text{else.} \end{cases}
\end{aligned}
$$

Table 4.1: Example for Dijkstra's algorithm, where v_4 is the initial vertex

It.nr	1		2		3		4		5		6	
Vertex	d_j	w_j	d_j	w_j	d_j	w_j	d_j	w_j	d_j	w_j	d_j	w_j
v_1	∞		∞		∞		6	3	6	3	6	3
v_2	∞		∞		∞		8	3	7	1	7	1
v_3	∞		4	6	4	6	4	6	4	6	4	6
v_4	0		0		0		0		0		0	
v_5	1	4	1	4	1	4	1	4	1	4	1	4
v_6	1	4	1	4	1	4	1	4	1	4	1	4
M_k	$\{1,2,3,5,6\}$		$\{1,2,3,5\}$		$\{1,2,3\}$		$\{1,2\}$		$\{2\}$		$\emptyset$	
v_i	4		6		5		3		1		2	

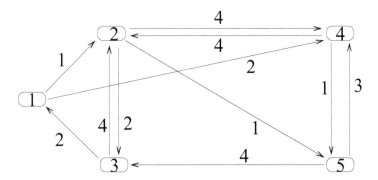

Figure 4.2: Sample graph for the Floyd–Warshall algorithm

Then, for $k = 1, ..., m$, where m is the number of vertices in V matrices $D_k = \{d_{ij}^{(k)}\}$ and $W_k = \{w_{ij}^{(k)}\}$ are computed as

$$
d_{ij}^{(k)} = \begin{cases} d_{ij}^{(k-1)} & \text{if} \quad i=j, i=k \text{ or } j=k, \\ \min\{d_{ij}^{(k-1)}, d_{ik}^{(k-1)} + d_{kj}^{(k-1)}\} & \text{else}, \end{cases}
$$
$$
w_{ij}^{(k)} = \begin{cases} w_{kj}^{(k-1)} & \text{if} \quad d_{ij}^{(k)} < d_{ij}^{(k-1)}, \\ w_{ij}^{(k-1)} & \text{else}. \end{cases}
$$

The value $d_{ij}^{(m)}$ represents the length or cost of the shortest path from v_i to v_j, while the value $w_{ij}^{(m)}$ denotes the direct ancestor of v_j along the shortest path from v_i to v_j. Using the matrix W_k it is easily possible to obtain the whole shortest path from a vertex v_i

to another vertex v_j in a recursive fashion. As a computational example consider the graph displayed in Figure 4.2. With respect to the matrices D_k and W_k for $k = 0, .., 5$ we obtain the following.

$$D_0 = \begin{bmatrix} 0 & 1 & \infty & 2 & \infty \\ \infty & 0 & 2 & 4 & 1 \\ 2 & 4 & 0 & \infty & \infty \\ \infty & 4 & \infty & 0 & 1 \\ \infty & \infty & 4 & 3 & 0 \end{bmatrix} \qquad W_0 = \begin{bmatrix} 1 & 1 & 0 & 1 & 0 \\ 0 & 2 & 2 & 2 & 2 \\ 3 & 3 & 3 & 0 & 0 \\ 0 & 4 & 0 & 4 & 4 \\ 0 & 0 & 5 & 5 & 5 \end{bmatrix}$$

$$D_1 = \begin{bmatrix} 0 & 1 & \infty & 2 & \infty \\ \infty & 0 & 2 & 4 & 1 \\ 2 & 3 & 0 & 4 & \infty \\ \infty & 4 & \infty & 0 & 1 \\ \infty & \infty & 4 & 3 & 0 \end{bmatrix} \qquad W_1 = \begin{bmatrix} 1 & 1 & 0 & 1 & 0 \\ 0 & 2 & 2 & 2 & 2 \\ 3 & 1 & 3 & 1 & 0 \\ 0 & 4 & 0 & 4 & 4 \\ 0 & 0 & 5 & 5 & 5 \end{bmatrix}$$

$$D_2 = \begin{bmatrix} 0 & 1 & 3 & 2 & 2 \\ \infty & 0 & 2 & 4 & 1 \\ 2 & 3 & 0 & 4 & 5 \\ \infty & 4 & 6 & 0 & 1 \\ \infty & \infty & 4 & 3 & 0 \end{bmatrix} \qquad W_2 = \begin{bmatrix} 1 & 1 & 2 & 1 & 2 \\ 0 & 2 & 2 & 2 & 2 \\ 3 & 1 & 3 & 1 & 2 \\ 0 & 4 & 2 & 4 & 4 \\ 0 & 0 & 5 & 5 & 5 \end{bmatrix}$$

$$D_3 = \begin{bmatrix} 0 & 1 & 3 & 2 & 2 \\ 4 & 0 & 2 & 4 & 1 \\ 2 & 3 & 0 & 4 & 5 \\ 8 & 4 & 6 & 0 & 1 \\ 6 & 7 & 4 & 3 & 0 \end{bmatrix} \qquad W_3 = \begin{bmatrix} 1 & 1 & 2 & 1 & 2 \\ 3 & 2 & 2 & 2 & 2 \\ 3 & 1 & 3 & 1 & 2 \\ 3 & 4 & 2 & 4 & 4 \\ 3 & 3 & 5 & 5 & 5 \end{bmatrix}$$

$$D_4 = \begin{bmatrix} 0 & 1 & 3 & 2 & 2 \\ 4 & 0 & 2 & 4 & 1 \\ 2 & 3 & 0 & 4 & 5 \\ 8 & 4 & 6 & 0 & 1 \\ 6 & 7 & 4 & 3 & 0 \end{bmatrix} \qquad W_4 = \begin{bmatrix} 1 & 1 & 2 & 1 & 2 \\ 3 & 2 & 2 & 2 & 2 \\ 3 & 1 & 3 & 1 & 2 \\ 3 & 4 & 2 & 4 & 4 \\ 3 & 3 & 5 & 5 & 5 \end{bmatrix}$$

$$D_5 = \begin{bmatrix} 0 & 1 & 3 & 2 & 2 \\ 4 & 0 & 2 & 4 & 1 \\ 2 & 3 & 0 & 4 & 5 \\ 7 & 4 & 5 & 0 & 1 \\ 6 & 7 & 4 & 3 & 0 \end{bmatrix} \qquad W_5 = \begin{bmatrix} 1 & 1 & 2 & 1 & 2 \\ 3 & 2 & 2 & 2 & 2 \\ 3 & 1 & 3 & 1 & 2 \\ 5 & 4 & 5 & 4 & 4 \\ 3 & 3 & 5 & 5 & 5 \end{bmatrix}$$

So, for example, the path from v_5 to v_2 has a length of 7 and passes through v_5, v_3, v_1, and v_2. Note that this is a good example for a non–unique shortest path since also the path from v_5 to v_2 that passes through v_4 has a length of 7.

4.2   Cost Analysis for Hierarchical Models Based on Poisson Line Processes

In this section we investigate a first scenario with respect to cost analysis. In particular, two–level hierarchical models based on two Cox point processes as defined in Section 2.4.6 are considered. Here, a characteristic that is of importance is the shortest path length, i.e., the distance measured along the lines of the underlying Poisson line process X_ℓ between a point of the lower–level point process X_L to its nearest (with respect to Euclidean distance) neighbour of the higher–level point process X_H. The results of Section 4.2 are partially based on results obtained in [33] and [34].

4.2.1   Model Definition

With respect to a stationary and isotropic Poisson line process X_ℓ with intensity γ consider a (non–marked) Cox point process (see Section 2.4.6) $X_H = \{X_n\}_{n\geq 1}$ with random driving measure

$$\Lambda_{X_H}(B) = \lambda_{H_\ell} \nu_1(B \cap X_\ell),$$

and finite intensity $\lambda_H = \lambda_{H_\ell}\gamma > 0$. Additionally, regard a second Cox point process $\widetilde{X}_L = \{\widetilde{X}_n\}_{n\geq 1}$ with random driving measure

$$\Lambda_{X_L}(B) = \lambda_{L_\ell} \nu_1(B \cap X_\ell),$$

and finite intensity $\lambda_L = \lambda_{L_\ell}\gamma > 0$. Note that, given X_ℓ, the two point processes X_H and $\widetilde{X}_L$ are considered to be independent. Furthermore, we denote by $N(\widetilde{X}_n)$ the location of the nearest (in the Euclidean sense) point of X_H with respect to the point $\widetilde{X}_n \in \widetilde{X}_L$. The point processes X_H and $\widetilde{X}_L$ then form a two–level hierarchical model (Figure 4.3). Note that although, with respect to the choice of the point processes that are based on the Poisson line process X_ℓ we restrict ourselves to Cox point processes in the sense of Section 2.4.6, some of the results can be extended to other types of (stationary and ergodic) point processes. In the following we suppose that each point X_n of X_H has an influence zone $\Xi(X_n)$ given by the Voronoi cell of X_n with respect to X_H. Thereby the sequence of influence zones $\{\Xi(X_n)\}_{m\geq 1}$ forms a Voronoi tessellation τ_X induced by X_H (Figure 4.3c).

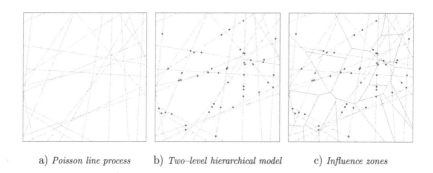

a) *Poisson line process*   b) *Two–level hierarchical model*   c) *Influence zones*

Figure 4.3: Two–level hierarchical model based on two Cox point processes X_H and X_L

Theorem 4.1 *Let Ξ_τ^* denote the typical cell of the Voronoi tessellation $\tau = \{\Xi(X_n)\}_{n\geq 1}$ induced by the stationary point process $X_H = \{X_n\}_{n\geq 1}$ of higher–level points. Then,*

$$\lambda_{H_\ell} = \frac{1}{\mathbb{E}_{X_H}\nu_1(L(\Xi_\tau^*))}, \tag{4.1}$$

where $\mathbb{E}_{X_H}$ denotes expectation with respect to the Palm probability measure $P_{X_H}^*$ introduced in (2.5) and where $L(\Xi_\tau^*)$ denotes the (Palm) line system within the typical cell Ξ_τ^*.

Proof   Using $\lambda_H = \lambda_{H_\ell}\gamma$ and (2.25), it becomes immediately clear that

$$\frac{1}{\lambda_{H_\ell}\gamma} = \mathbb{E}_{X_H}\nu_2(\Xi_\tau^*).$$

Furthermore, we have $\mathbb{E}_{X_H}(\nu_1(L(\Xi_\tau^*))) = \gamma\mathbb{E}_{X_H}\nu_2(\Xi_\tau^*)$ (cmp. [34]). This proves (4.1).   □

4.2.2   Transformation to Graph Structure

The two–level hierarchical model based on a common underlying Poisson line process X_ℓ described in Section 4.2 can also be regarded as a graph structure (Figure 4.4). For this purpose we regard the segments or edges that are formed by the lines of the Poisson line process X_ℓ. Points of X_H and X_L are considered as nodes of the graph. If an edge contains a node of the graph the edge is subdivided into two edges. Now, a cost or weight $c(e)$ is assigned to each edge e, e.g., the length of e. By introducing an additional connection rule between the nodes of the graph inference, for example, about the average distance between a node of the lower–level to its nearest neighbouring higher–level node can be obtained.

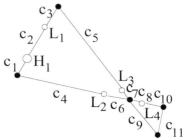

a) *Two–level hierarchical model*

b) *Corresponding graph structure with higher–level node H_1, lower–level nodes L_i and edge costs c_i*

Figure 4.4: Transformation of two–level hierarchical model into graph structure

4.2.3   Mean Shortest Path Lengths and Mean Subscriber Line Lengths

For the two–level hierarchical model defined in Section 4.2.1 a cost functional of interest is the *shortest path length*, in particular the *mean shortest path length* c_{sp} that is defined as the mean distance from a point of the lower–level Cox point process $\widetilde{X}_L$ to the nearest (in the Euclidean sense) point belonging to the higher–level Cox point process X_H. This distance is measured along the lines of the underlying Poisson line process X_ℓ, or in other words along the edges of the transformed graph structure (cmp. Section 4.2.2). So, in order to analyse shortest path lengths, each location $\widetilde{X}_n$ of $\widetilde{X}_L'$ is associated with the mark $c(P(N(\widetilde{X}_n), \widetilde{X}_n)) > 0$ representing the length of the shortest path $P(N(\widetilde{X}_n), \widetilde{X}_n)$ between the location of a lower–level point $\widetilde{X}_n$ and its nearest higher–level point $N(\widetilde{X}_n)$. Thereby a stationary marked point process

$$X_L = \{[\widetilde{X}_n, c(P(\widetilde{X}_n, N(\widetilde{X}_n)))]\}_{n\geq 1}\,, \tag{4.2}$$

can be constructed whose mark space is the non–negative x–axis. The mean shortest path length c_{sp} is then the average with respect to the Palm mark distribution of X_L. Therefore, due to the stationarity of X_L, we are able to express c_{sp} by

$$c_{sp} = \frac{1}{\lambda_L \nu_2(B)} \,\mathbb{E} \sum_{n\geq 1} \mathbb{I}_B(\widetilde{X}_n) c(P(\widetilde{X}_n, N(\widetilde{X}_n))) = \mathbb{E}_{X_L} c(P(o, N(o)))\,. \tag{4.3}$$

Recall that the symbol B in (4.3) means an arbitrary (bounded) Borel set $B \in \mathcal{B}(\mathbb{R}^2)$ with $0 < \nu_2(B) < \infty$ and $\mathbb{E}_{X_L}$ denotes expectation with respect to the Palm probability measure $\mathbb{P}^*_{X_L}$ introduced in (2.20).

Note that, by using the ergodicity of the point process X_L, it is possible to represent the mean shortest path length c_{sp} in an alternative fashion as the limit of a spatial average. For this purpose we define $c_{sp}(W)$ as the average shortest path length in a sampling window W, where

$$c_{sp}(W) = \frac{1}{\#\{n : \widetilde{X}_n \in W\}} \sum_{n \geq 1} \mathbb{1}_W(\widetilde{X}_n) c(P(\widetilde{X}_n, N(\widetilde{X}_n))) . \tag{4.4}$$

If we consider an averaging sequence $\{W_i\}_{i \geq 1}$ of bounded sampling windows (cmp. Section 2.1) then we get that (cmp. [93])

$$\lim_{i \to \infty} c_{sp}(W_i) = c_{sp} \tag{4.5}$$

holds with probability one, where c_{sp} is given by (4.3). Note that, with respect to the estimation of c_{sp}, edge effects can occur, e.g. if $\widetilde{X}_n \in W$ but $N(\widetilde{X}_n) \notin W$ and that it might be hard to come up with an edge–corrected estimator.

A slight modification of the mean shortest path length c_{sp} is given by the *mean subscriber line length* c_{sl}. Here, the locations of the lower–level points are not distributed according to a Cox point process $\widetilde{X}_L$ with random intensity measure Λ_{X_L} described in Section 4.2.1 but according to a (stationary) Poisson point process $\widetilde{X}'_L$ with (planar) intensity λ'_L that is independent of X_ℓ and X_H. As in the scenario for the mean shortest path length a lower–level point $\widetilde{X}'_n \in \widetilde{X}'_L$ is connected to its nearest (in the Euclidean sense) point $N(\widetilde{X}'_n) \in X_H$. For this purpose we first connect the point $\widetilde{X}'_n$ to its nearest point $\widetilde{X}''_n$ of the line system $L(\Xi_n)$ (Figure 4.5), where $\Xi_n = \Xi(N(\widetilde{X}'_n))$ is the Voronoi cell of $N(\widetilde{X}'_n)$ and $L(\Xi_n)$ is given by the restriction of the Poisson line process X_ℓ to Ξ_n. The distance $c(P(\widetilde{X}'_n, N(\widetilde{X}'_n)))$ from the lower–level point $\widetilde{X}'_L$ to $N(\widetilde{X}'_n) \in X_H$ can then be expressed by

$$c(P(\widetilde{X}'_n, N(\widetilde{X}'_n))) = c'(\widetilde{X}'_n, \widetilde{X}''_n) + c(P(\widetilde{X}''_n, N(\widetilde{X}'_n))) , \tag{4.6}$$

where $c'(\widetilde{X}'_n, \widetilde{X}''_n)$ is considered to be the cost value of the (virtual) edge with respective endpoints $\widetilde{X}'_n$ and $\widetilde{X}''_n$. Note that in the context of telecommunication the (virtual) edge between a lower–level point $\widetilde{X}'_n$ and its projected point $\widetilde{X}''_n \in L(\Xi_n)$ is often called the *last meter*. With respect to the results of Monte–Carlo simulations presented in Section 4.2.8 we take $c'(\widetilde{X}'_n, \widetilde{X}''_n) = 0$ but note that other choices are also possible, e.g. the Euclidean distance.

The mean subscriber line length c_{sl} can now be defined as

$$c_{sl} = \frac{1}{\lambda_L \nu_2(B)} \mathbb{E} \sum_{n \geq 1} \mathbb{1}_B(X'_n) c(P(\widetilde{X}'_n, N(\widetilde{X}'_n))) = \mathbb{E}_{X'_L} c(P(o, N(o))) \tag{4.7}$$

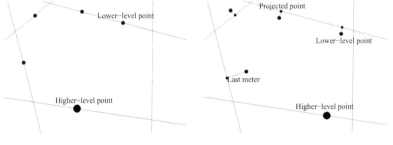

a) *Linear placement on lines (mean shortest   b) Spatial placement and projection to nearest
path length c_{sp})*   *line (mean subscriber line length c_{sl})*

Figure 4.5: Two scenarios for the two–level hierarchical model

for some (bounded) Borel set $B \in \mathcal{B}(\mathbb{R}^2)$ with $0 < \nu_2(B) < \infty$, where the cost value $c(P(\widetilde{X}'_n, N(\widetilde{X}'_n)))$ of the shortest path from $\widetilde{X}'_n$ to $N(\widetilde{X}'_n)$ is given in (4.6).

Similar to the case of the mean shortest path length c_{sp}, we are able to provide an alternative representation of the mean subscriber line length c_{sl} by utilizing the ergodicitiy of X'_L. Let $c_{sl}(W)$ be the mean subscriber line length with respect to a sampling window $W \subset \mathbb{R}^2$ being defined by

$$c_{sl}(W) = \frac{1}{\#\{n : \widetilde{X}'_n \in W\}} \sum_{n \geq 1} \mathbb{1}_W(\widetilde{X}'_n) \, c(P(\widetilde{X}'_n, N(\widetilde{X}'_n))). \qquad (4.8)$$

By the ergodicity of X'_L we obtain that (cmp. [93])

$$\lim_{i \to \infty} c_{sl}(W_i) = c_{sl}. \qquad (4.9)$$

if $\{W_i\}_{i \geq 1}$ is an averaging sequence of bounded sampling windows (cmp. Section 2.1). Note that the estimation of $c_{sl}(W)$ is hindered by occuring edge effects and that therefore in the following an alternative representation as well as an alternative estimator is derived.

4.2.4   Application of Neveu's Formula

Regarding the alternative representations for the mean shortest path length c_{sp} as well as the mean subscriber line length c_{sl} that are provided in (4.5) and (4.9) a natural way of estimating these two characteristics is induced. In a possibly large sampling window

W the processes X_ℓ, X_H and $\widetilde{X}_L$ or $\widetilde{X}'_L$, respectively, are realised and the mean over all distances between the locations of lower–level points to their nearest higher–level point is taken as an estimator for c_{sp} or c_{sl}, respectively. However, it turns out that by an application of Neveu's exchange formula given in (2.21) alternative estimators for c_{sp} and c_{sl} can be derived that completely avoid the simulation of lower–level locations. Such efficient estimators for c_{sp} and c_{sl} are based on the following two theorems.

Theorem 4.2 *Consider the point process $X_H = \{X_n\}_{n\geq 1}$ of higher–level points and the (marked) point process $X_L = \{[\widetilde{X}_n, c(P(\widetilde{X}_n, N(\widetilde{X}_n)))]\}_{n\geq 1}$ defined in (4.2). Then,*

$$\mathbb{E}_{X_L}\, c(P(o, N(o))) = \frac{1}{\mathbb{E}_{X_H} \nu_1(L(\Xi^*_\tau))}\, \mathbb{E}_{X_H} \int_{L(\Xi^*_\tau)} c(P(u, o))\, du\,, \qquad (4.10)$$

where Ξ^*_τ denotes the typical cell of the Voronoi tessellation τ induced by X_H and $L(\Xi^*_\tau)$ is the (Palm) line system within Ξ^*_τ.

Proof   The proof of Theorem 4.2 is based on Neveu's exchange formula (see (2.21)) for jointly stationary point processes, which are defined on a common probability space $(\Omega, \mathcal{A}, \mathbb{P})$ equipped with some flow $\{\theta_x,\ x \in \mathbb{R}^2\}$. We use (2.21) with X_D and $\widetilde{X}_{\widetilde{D}}$ being equal to X_H and X_L, respectively. Thus, the mark space $\mathbb{M}$ will be omitted and $\widetilde{\mathbb{M}} = [0, \infty)$. Recall that both intensities λ_H and λ_L of X_H and X_L, respectively, can be expressed by λ_{H_ℓ}, λ_{L_ℓ}, and γ (cmp. Section 4.2.1). In particular,

$$\lambda_H = \lambda_{H_\ell} \gamma \qquad \text{and} \qquad \lambda_L = \lambda_{L_\ell} \gamma\,. \qquad (4.11)$$

Additionally, we consider the function $f : \mathbb{R}^2 \times [0, \infty) \times \Omega \to [0, \infty)$ that is given by

$$f(x, \widetilde{g}, \omega) = \begin{cases} \widetilde{g} & \text{if } X_H(\theta_{-x}\omega, B^{\neq}_{|x|}(x)) = 0\,, \\ 0 & \text{otherwise} \end{cases} \qquad (4.12)$$

for any $x \in \mathbb{R}^2$, $\widetilde{g} = c(P(x, o)) \geq 0$, and $\omega \in \Omega$, where $B^{\neq}_{|x|}(x) = \{y \in \mathbb{R}^2 : |y - x| < |x|\}$. Then, $f(x, \widetilde{g}, \omega) = \widetilde{g}$ if $-x \in \mathbb{R}^2$ is an atom of the counting measure $X_H(\omega, \cdot)$ such that there are no other atoms of $X_H(\omega, \cdot)$ which are closer (in the Euclidean sense) to the origin o than $-x$. Hence, applying Neveu's exchange formula (2.21), we obtain that

$$\mathbb{E}_{X_L} c(P(o, N(o))) = \int_{\Omega \times \mathbb{M}} \int_{\mathbb{R}^2} f(-x, \widetilde{g}, \omega) X_H(\omega, dx) P^*_{X_L}(d(\omega, \widetilde{g}))$$

$$= \frac{\lambda_H}{\lambda_L} \int_\Omega \int_{\mathbb{R}^2 \times \mathbb{M}} f(x, \widetilde{g}, \theta_x \omega) X_L(\omega, d(x, \widetilde{g})) P^*_{X_H}(d\omega)\,. \qquad (4.13)$$

Note that, given the typical Voronoi cell Ξ_τ^* and the (typical) line system $L(\Xi_\tau^*)$ within Ξ_τ^*, the (random) number of points of X_L on $L(\Xi_\tau^*)$ is Poisson distributed with expectation $\eta = \lambda_{L_\ell}\nu_1(L(\Xi_\tau^*))$. Thus, taking into account the definition of the function f given in (4.12), the inner integral on the right hand side of (4.13) can be expressed as

$$\int_{\mathbb{R}^2\times\mathbb{M}} f(x,\widetilde{g},\theta_x\omega)X_L(\omega,d(x,\widetilde{g})) = \sum_{k=1}^{\infty} e^{-\eta}\frac{\eta^k}{k!} \int_{L(\Xi_\tau^*)}\cdots\int_{L(\Xi_\tau^*)} \sum_{i=1}^{k} \frac{c(P(u_i,o))}{\nu_1(L(\Xi_\tau^*))^k}\,du_1\ldots du_k\,,$$

due to the conditional uniform distribution of the lower–level points on $L(\Xi_\tau^*)$. Thus,

$$\int_{\mathbb{R}^2\times\mathbb{M}} f(x,\widetilde{g},\theta_x\omega)X_L(\omega,d(x,\widetilde{g})) = \sum_{k=1}^{\infty} e^{-\eta}\frac{\eta^k}{k!}\frac{k}{\nu_1(L(\Xi_\tau^*))} \int_{L(\Xi_\tau^*)} c(P(u,o))\,du$$

$$= \lambda_{L_\ell} \int_{L(\Xi_\tau^*)} c(P(u,o))\,du\,.$$

Summarizing things, we get that

$$\mathbb{E}_{X_L}c(P(o,N(o))) = \frac{\lambda_H}{\lambda_L}\lambda_{L_\ell}\,\mathbb{E}_{X_H} \int_{L(\Xi_\tau^*)} c(P(u,o))\,du\,.$$

Combining this with (4.1) and (4.11) completes the proof of the theorem. $\qquad\square$

Analogously to Theorem 4.2 for the case of the mean shortest path length c_{sp} the following theorem can be given for the case of the mean subscriber line length c_{sl}.

Theorem 4.3 *Consider the point process $X_H = \{X_n\}_{n\geq1}$ of higher–level points and the (marked) point process $X_L' = \{[\widetilde{X}_n', c(P(\widetilde{X}_n', N(\widetilde{X}_n')))]\}_{n\geq1}$ constructed by the sequence of lower–level points $\{\widetilde{X}_n'\}_{n\geq1}$ and the marks $c(P(\widetilde{X}_n', N(\widetilde{X}_n')))$ defined in (4.6). Then,*

$$\mathbb{E}_{X_L'}\,c(P(o,N(o))) = \frac{1}{\mathbb{E}_{X_H}\nu_2(\Xi_\tau^*)}\,\mathbb{E}_{X_H} \int_{\Xi_\tau^*} c(P(u,o))\,du\,, \qquad (4.14)$$

where Ξ_τ^* denotes the typical cell of the Voronoi tessellation τ induced by X_H.

Proof   Using Neveu's exchange formula (2.21) and proceeding similarly as in the proof of Theorem 4.2, we obtain that

$$\mathbb{E}_{X_L'}c(P(o,N(o))) = \frac{\lambda_H}{\lambda_L}\lambda_L\,\mathbb{E}_{X_H} \int_{\Xi_\tau^*} c(P(u,o))\,du = \lambda_H\,\mathbb{E}_{X_H} \int_{\Xi_\tau^*} c(P(u,o))\,du\,,$$

where in the first equality it is used that, given the typical Voronoi cell Ξ_τ^* and the (typical) line system $L(\Xi_\tau^*)$ restricted to Ξ_τ^*, the random number of points of X_L' within

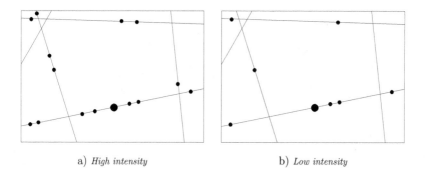

a) *High intensity*   b) *Low intensity*

Figure 4.6: Different intensities of lower–level points but same mean distance to the nearest higher–level point

Ξ_τ^* is Poisson distributed with expectation $\eta' = \lambda_L \nu_2(\Xi_\tau^*)$. The proof is then finished by using that $\lambda_H^{-1} = \mathbb{E}_{X_H} \nu_2(\Xi_\tau^*)$. $\square$

Note that by using Theorems 4.2 and 4.3 it is possible to simplify the computation of the mean shortest path length c_{sp} and the mean subscriber line length c_{sl}, respectively, by estimating the quotients of expectations appearing on the right–hand sides of (4.10) and (4.14), respectively. In order to estimate these quotients the typical serving zone Ξ_τ^* of higher–level points has to be simulated, together with their corresponding (typical) line system, where $L(\Xi_\tau^*)$ denotes this line system restricted to Ξ_τ^*. A simulation algorithm that is capable of performing such a simulation is explained in Section 3.2.2 since the typical serving zone Ξ_τ^* coincides with the typical cell of the Cox–Voronoi tessellation τ induced by a Cox point process X_c given in Section 2.4.6. It should also be remarked that the expression for $\mathbb{E}_{X_L} c(P(o, N(o)))$ given in (4.10) can alternatively be written in the form

$$\mathbb{E}_{X_L} c(P(o, N(o))) = \lambda_{H_\ell} \mathbb{E}_{X_H} \int_{L(\Xi_\tau^*)} c(P(u, o)) \, du. \qquad (4.15)$$

This is an immediate corollary from Theorems 4.1 and 4.2. In particular this representation of c_{sp} shows that it does not depend on λ_{L_ℓ} (cmp. Figure 4.6).

4.2.5 Efficient Estimation of the Mean Shortest Path Length

In order to derive an efficient estimator $\widehat{c}_{sp}$ for the mean shortest path length c_{sp}, we can use the representation of c_{sp} given in Theorem 4.2. Thus, the basic idea is a simulation

of the typical Voronoi cell Ξ_τ^* and of the (typical) line system $L(\Xi_\tau^*)$ a certain number
of times, n say. Additionally, we partition the line system $L(\Xi_{\tau_i}^*)$ in $\Xi_{\tau_i}^*$ for $i = 1, ..., n$
into its line segments $E_i = \{S_i^{(1)}, S_i^{(2)} ..., S_i^{(M_i)}\}$, where M_i is the total number of line
segments in $\Xi_{\tau_i}^*$ for $1 \leq i \leq n$. Note that the line segment containing the origin (and
therefore the higher–level point) is subdivided into two segments (Figure 4.7a).

Taking a classical sample mean it is obtained that $\lim_{n\to\infty} \widehat{c}_{sp}(n) = c_{sp}$ with probability
one, where

$$\widehat{c}_{sp}(n) \quad = \quad \frac{1}{\frac{1}{n}\sum_{i=1}^{n} \nu_1(L(\Xi_i^*))} \frac{1}{n} \sum_{i=1}^{n}\sum_{j=1}^{M_i} \int_{S_i^{(j)}} c(P(u,o))\, du, \tag{4.16}$$

$$= \quad \frac{1}{\sum_{i=1}^{n} \nu_1(L(\Xi_i^*))} \sum_{i=1}^{n}\sum_{j=1}^{M_i} \int_{S_i^{(j)}} c(P(u,o))\, du. \tag{4.17}$$

Note that if the (linear) intensity λ_{H_ℓ} is known then, by using the relationship (4.15)
an alternative estimator $\check{c}_{sp}(n)$ for c_{sp} can be derived, where

$$\check{c}_{sp}(n) = \lambda_{H_\ell} \frac{1}{n} \sum_{i=1}^{n}\sum_{j=1}^{M_i} \int_{S_i^{(j)}} c(P(u,o))\, du. \tag{4.18}$$

For both estimators $\widehat{c}_{sp}(n)$ and $\check{c}_{sp}(n)$ a question that arises is how to compute the
integrals appearing on the righ–hand sides of (4.17) and (4.18), respectively. This
question is answered by the following theorem, where some additional assumptions are
made on the cost function $c : E \to [0, \infty)$ introduced in Section 4.1.

Theorem 4.4 *Let the values $c(e)$ of the cost function $c : E \to [0, \infty)$ only depend on
the lengths of the edges $e \in E$ and suppose that $c(e)$ is monotonously increasing with
respect to the length of e, where $c(e) = 0$ if $\nu_1(e) = 0$. Furthermore, let $S = S(A, B)$ be
a line segment with respective endpoints A and B, and let $\delta_S = c(P(B, o)) - c(P(A, o))$.
Then it holds that*

$$c(P(A, B)) \geq |\delta_S|. \tag{4.19}$$

and that there exists a point $D \in S$ such that

$$c(P(A, o)) + c(P(D, A)) = c(P(B, o)) + c(P(D, B)) \tag{4.20}$$

and

$$\int_S c(P(u, o))\, du \quad = \quad c(P(A, o))\nu_1(D - A) + \int_D^A c(P(A, u))\, du$$

$$+ c(P(B, o))\nu_1(D - B) + \int_D^B c(P(B, u))\, du. \tag{4.21}$$

Proof First it is shown that (4.19) holds. If $c(P(B,o)) = c(P(A,o))$, we obviously have that $c(P(A,B)) \geq |\delta_S| = 0$. Now let $c(P(B,o)) > c(P(A,o))$ and suppose that

$$c(P(B,o)) > c(P(A,B)) + c(P(A,o)).$$

Then the path length $c(P(A,B)) + c(P(A,o))$ from B to o passing through A would be smaller than $c(P(B,o))$, which is a contradiction to the definition of the shortest path length $c(P(B,o))$. Therefore, (4.19) is shown. If $c(P(A,B)) = 0$, then, by using the monotonicity of $c : E \to [0,\infty)$, we obtain that

$$c(P(B,u)) = c(P(A,u))$$

for each $u \in S$. Furthermore, (4.19) implies that

$$c(P(B,o)) = c(P(A,o)),$$

and this means that (4.20) and (4.21) are obviously true for any $D \in S$. We now assume that $c(P(A,B)) > 0$. First suppose that

$$c(P(B,o)) = c(P(A,B)) + c(P(A,o)).$$

It is easy to see that (4.20) and (4.21) hold for $D = B$. If $c(P(B,o)) > c(P(A,o))$ and

$$c(P(B,o)) < c(P(A,B)) + c(P(A,o)),$$

then, by the monotonicity of $c : E \to [0,\infty)$, there is a distance peak D which lies between the two endpoints A and B, respectively, of S (see Figure 4.7b). Note that the distance peak D can be defined as an inner point of the segment S, where $c(P(D,o))$ takes the same value no matter if the origin o is reached passing through A or passing through B. In other words, (4.20) and (4.21) hold. □

For the special case of $c(S)$ being the length of the segment S the following corollary can be provided that facilitates the computation of the integral appearing on the left hand side of (4.21).

Corollary 4.1 *Let $c(S)$ be the length of the segment $S = S(A,B)$, that means $c(S) = \nu_1(S)$, then*

$$\int_S c(P(u,o)) \, du = f(\nu_1(S); c(P(A(S),o)), c(P(B(S),o))), \tag{4.22}$$

where

$$f(x; \theta_1, \theta_2) = \frac{1}{4}x^2 + \frac{1}{2}(\theta_1 + \theta_2)x - \frac{1}{4}(\theta_2 - \theta_1)^2. \tag{4.23}$$

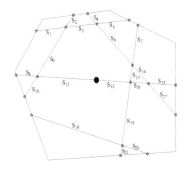 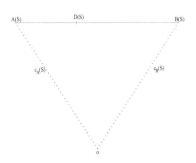

a) *Partitioning of a realisation of $L(\Xi_\tau^*)$ into*   b) *Mean shortest path length for a single*
segments   *segment*

Figure 4.7: Partitioning and weighted mean shortest path length

Proof   Utilizing the abbreviations $c_A(S) = c(P(A, o))$ and $c_B(S) = c(P(B, o))$ as well
as $\delta_S = c_B(S) - c_A(S)$ (defined in Theorem 4.4) and bearing in mind that $c(S) = \nu_1(S)$,
(4.20) induces that

$$\nu_1(D - A) = \frac{\nu_1(S) + c_B(S) - c_A(S)}{2},$$

and that

$$\nu_1(D - B) = \frac{\nu_1(S) + c_A(S) - c_B(S)}{2}.$$

Additionally, we can write (4.21) in the form

$$\int_S c(P(u, o))\, du = f_1(\nu_1(S); c_A(S), c_B(S)) + f_2(\nu_1(S); c_A(S), c_B(S)),$$

where

$$f_1(x; \theta_1, \theta_2) = \frac{x + \theta_2 - \theta_1}{2}\, \frac{2\theta_1 + 1/2(x + \theta_2 - \theta_1))}{2} \qquad (4.24)$$

and

$$f_2(x; \theta_1, \theta_2) = \frac{x + \theta_1 - \theta_2}{2}\, \frac{\theta_1 + \theta_2 + 1/2(x + \theta_2 - \theta_1))}{2}. \qquad (4.25)$$

Note that $f_1(\nu_1(S); c_A(S), c_B(S))$ represents the sum of the first two summands in
(4.21), whereas $f_2(\nu_1(S); c_A(S), c_B(S))$ represents the sum of the last two summands in
(4.21). By simple calculations, we now get that the sum of the expressions in (4.24)
and (4.25) gives (4.23).   □

By Corollary 4.1, we immediately obtain the following expressions for the estimators $\widehat{c}_{sp}(n)$ and $\check{c}_{sp}(n)$ under the assumption that $c(S)$ is given by $\nu_1(S)$.

Corollary 4.2 *For each $n \geq 1$ let $E_i = \{S_i^{(j)}\}_{j=1}^{M_i}$ be the partition of the line system $L(\Xi_{\tau_i}^*)$ restricted to the ith typical cell $\Xi_{\tau_i}^*$ for $i = 1, \ldots, n$ and let $A_i^{(j)}$ and $B_i^{(j)}$, respectively, denote the endpoints of the segment $S_i^{(j)}$. Then,*

$$\widehat{c}_{sp}(n) = \frac{1}{\sum_{i=1}^{n} \nu_1(L(\Xi_{\tau_i}^*))} \sum_{i=1}^{n} \sum_{j=1}^{M_i} f(\nu_1(S_i^{(j)}); c(P(A_i^{(j)}), o), c(P(B_i^{(j)}, o))) \qquad (4.26)$$

and

$$\check{c}_{sp}(n) = \frac{\lambda_{H_\ell}}{n} \sum_{i=1}^{n} \sum_{j=1}^{M_i} f(\nu_1(S_i^{(j)}); c(P(A_i^{(j)}), o), c(P(B_i^{(j)}, o))), \qquad (4.27)$$

where the function f is given in (4.23).

Applying the representation formulae (4.26) and (4.27) it is sufficient to compute path lengths $c(P(A_i^{(j)}), o)$ and $c(P(B_i^{(j)}), o)$ for $j = 1, \ldots, M_i$ and $i = 1, \ldots, k$ for the determination of the estimators $\widehat{c}_{sp}(k)$ and $\check{c}_{sp}(k)$. Herefore, for example, Dijkstra's algorithm explained in Section 4.1.2 can be used.

4.2.6 Efficient Estimation of the Mean Subscriber Line Length

Using (4.7) and (4.14), it is possible to derive an estimator $\widehat{c}_{sl}$ for the mean subscriber line length c_{sl} considered in (4.8). As in the case for the estimation of the mean shortest path length c_{sp} we start by simulating the typical Voronoi cell Ξ_τ^* and the typical line system both with respect to the higher–level points n times, obtaining n independent and identically distributed copies $\Xi_{\tau,1}^*, \ldots, \Xi_{\tau,n}^*$ of Ξ_τ^*, where $n > 0$ is an arbitrary and fixed integer. In Figure 4.8 a sample for the typical Voronoi cell and the corresponding line system restricted to the typical cell is displayed. Apart from the nucleus of the typical Voronoi cell (thick gray dot), the typical Voronoi cell itself (blue line segments), the underlying line system restricted to the typical Voronoi cell (thick red line segments), and the inner Voronoi tessellation with respect to the restricted underlying line system (thin red line segments) are displayed. The cells of the inner Voronoi tessellation are induced by the set of edges for the cells of the underlying line system as described in Section 2.4.4. Note however that the inner Voronoi cells are not formed with respect to the boundary of the typical Voronoi cell Ξ_τ^* due to the fact that a lower–level point located in Ξ_τ^* is projected solely to a segment of the underlying line system $L(\Xi_\tau^*)$ restricted to Ξ_τ^*. For example, in the realisation of the typical Voronoi cell Ξ_τ^* shown

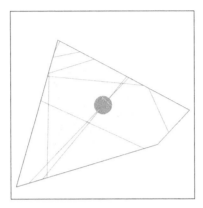

Figure 4.8: Sample of the typical Voronoi cell Ξ_τ^* with restricted underlying line structure $L(\Xi_\tau^*)$ and inner Voronoi tessellation with respect to $L(\Xi_\tau^*)$

in Figure 4.8, all points of the small triangle in the right corner are projected onto an isolated segment of the restricted line system $L(\Xi_\tau^*)$. This means that the shortest path from these points to the corresponding higher–level point, located at the origin, is not completely contained in the typical Voronoi cell Ξ_τ^*. The inner–Voronoi tessellation partitions the typical cell into a random number K of micro–cells $\{\Upsilon^{(j)}\}_{j=1,\dots,K}$, where each of these micro–cells Υ corresponds to a segment $S = S(A, B)$ of the line system $L(\Xi_\tau^*)$ restricted to Ξ_τ^*. The segment S is always an edge of two micro–cells. Each micro–cell Υ can further be partitioned into three non–overlapping subsets Υ_A, Υ_B, and Υ_C, respectively, as shown in Figure 4.9. In this example, we have that $B(S) = \Upsilon_B$ and the thick line on the left side is part of the boundary of the typical Voronoi cell Ξ_τ^*. Lower–level points that are located in Υ_B are connected to the endpoint $B(S)$, whereas the locations in Υ_C are projected onto S. Note that if the origin o (the higher–level point) belongs to the interior of the line segment S, then S is decomposed into two subsegments (A, o) and (B, o) and Υ_C is split into two sets corresponding to these two subsegments. For each $i = 1, \dots, n$, we have that the inner Voronoi tessellation completely partitions the ith copy Ξ_i^* of the typical cell into the micro–cells $\{\Upsilon_i^{(j)}\}_{j=1,\dots,K_i}$. Therefore, by taking classical sample means it is obtained that $\lim_{n\to\infty} \widehat{c}_{sl}(n) = c_{sl}$ with probability one, where

$$\widehat{c}_{sl}(n) = \frac{1}{\sum_{i=1}^n \nu_2(\Xi_{\tau_i}^*)} \sum_{i=1}^n \sum_{j=1}^{K_i} \int_{\Upsilon_i^{(j)}} c(P(u, o))\, du\,. \tag{4.28}$$

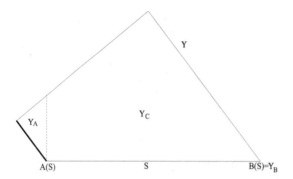

Figure 4.9: Partition of the micro–cell Υ into three subsets Υ_A, Υ_B, and Υ_C

Recall that the expression for $\mathbb{E}_{X'_L} c(P(o, N(o)))$ provided in (4.14) has an alternative representation of the form

$$\mathbb{E}_{X'_L} c(P(o, N(o))) = \lambda_{H_\ell} \gamma \, \mathbb{E}_{X_H} \int_{\Xi^*_\tau} c(P(u, o)) \, du \,. \tag{4.29}$$

Hence, for λ_{H_ℓ} and γ known, an alternative estimator $\check{c}_{sl}$ for c_{sl} is given by

$$\check{c}_{sl}(n) = \lambda_{H_\ell} \gamma \, \frac{1}{n} \sum_{i=1}^{n} \sum_{j=1}^{K_i} \int_{\Upsilon_i^{(j)}} c(P(u, o)) \, du \,. \tag{4.30}$$

Suppose now for the remaining part of this section in order to simplify the computations, that the cost value $c'(X'_n, X''_n)$ of the edge with respective endpoints X'_n and X''_n equals zero. Note that with some slight modifications it is possible to account for some other choices of $c'(X'_n, X''_n)$ like the Euclidean distance. Obviously, by (4.6), we have that

$$c(P(X'_n, N(X'_n))) = c(P(X''_n, N(X'_n))) \,.$$

Additionally, for the integral appearing in (4.28) and (4.30), we obtain that

$$\int_{\Upsilon_i^{(j)}} c(P(u, o)) \, du = \int_{\Upsilon_i^{(j)}} c(P(u_p, o)) \, du \,,$$

where u_p denotes the closest point, with respect to u, of the line system $L(\Xi^*_{\tau_i})$ within the set $\Upsilon_i^{(j)}$. The following theorem displays a way in order to compute this integral analytically.

Theorem 4.5 *Let Υ be a micro–cell within the typical cell Ξ_τ^* and let $S = S(A, B)$ be the corresponding segment with endpoints A and B, respectively, of the underlying line system $L(\Xi_\tau^*)$ restricted to Ξ_τ^*. Then, with the abbreviation $c_A(S) = c(P(A, o))$ and $c_B(S) = c(P(B, o))$,*

$$\int_\Upsilon c(P(u, o))\, du = c_A(S)\nu_2(\Upsilon_A) + c_B(S)\nu_2(\Upsilon_B) + \int_{\Upsilon_C} c(P(u, o))\, du\,. \qquad (4.31)$$

Theorem 4.5 immediately follows from additivity of the Lebesgue integral with respect to the integration domain.

It is important to mention that the first two summands on the right–hand side of (4.31) can be easily computed if Dijkstra's algorithm is used to determine $c_A(S)$ and $c_B(S)$, respectively (cmp. Section 4.2.8). Considering the computation of the third summand of (4.31), it is possible to proceed similarly as in the proof of formula (4.21) derived in Theorem 4.4. However, due to the necessity of subdividing Υ_C in order to obtain linear functions as integrands, the computation might become a little more challenging but no principle problems occur.

4.2.7   Scaling Invariance Properties

In order to allow for an efficient analysis of the mean shortest path lengths and the mean subscriber line lengths some scaling invariance properties have to be discussed first. Recall that we assume X_H to be a Cox point process as defined in Section 4.2.1 and that the whole two–level hierarchical model can be completely described by the three parameters λ_L, λ_{H_ℓ} and γ, where $\lambda_H = \gamma\lambda_{H_\ell}$ is the intensity of X_H and λ_L denotes the intensity of the lower level point process $\widetilde{X}'_L$ (either a Cox point process or a Poisson point process). Besides, it is assumed that $c(S)$ is the length of the segment S, i.e., $c(S) = \nu_1(S)$. In Section 4.2.4 it has been shown that both characteristics, the mean shortest path lengths c_{sp} as well as the mean subscriber line lengths c_{sl} are independent of the model parameter λ_L. Furthermore, it has been explained in Section 2.4.6 for characteristics of the typical cell that with respect to the two remaining parameters λ_{H_ℓ} and γ, a scaling invariance property holds for any fixed value of the quotient $\kappa = \gamma/\lambda_{H_\ell}$. Now, we derive a similar scaling invariance effect for c_{sp} and c_{sl}.

The following theorem shows that it is possible to provide estimates for the mean shortest path length as well as the mean subscriber line length with respect to a given parameter pair $(\gamma, \lambda_{H_\ell})$ by using estimates for a different parameter pair that has the same quotient κ and by performing a suitable standardisation afterwards.

Theorem 4.6 *For any pair* $(\gamma, \lambda_{H_\ell})$ *of parameters* $\gamma, \lambda_{H_\ell} > 0$, *consider the characteristics* $c_{sp} = c_{sp}(\gamma, \lambda_{H_\ell})$ *and* $c_{sl} = c_{sl}(\gamma, \lambda_{H_\ell})$ *given in* (4.3) *and* (4.7), *respectively. Then*

$$\gamma^{(1)} c_{sp}(\gamma^{(1)}, \lambda_{H_\ell}^{(1)}) = \gamma^{(2)} c_{sp}(\gamma^{(2)}, \lambda_{H_\ell}^{(2)}) \tag{4.32}$$

and

$$\gamma^{(1)} c_{sl}(\gamma^{(1)}, \lambda_{H_\ell}^{(1)}) = \gamma^{(2)} c_{sl}(\gamma^{(2)}, \lambda_{H_\ell}^{(2)}) \tag{4.33}$$

provided that $\gamma^{(1)}/\lambda_{H_\ell}^{(1)} = \gamma^{(2)}/\lambda_{H_\ell}^{(2)}$.

Proof It is only shown that (4.32) holds, because the proof of (4.33) is analogous. Let $\gamma^{(2)} = a\,\gamma^{(1)}$ and $\lambda_{H_\ell}^{(2)} = a\,\lambda_{H_\ell}^{(1)}$ for some $a > 0$. Then, we can use the scaling properties of the typical cell $\Xi_\tau^{*(i)}$ of the Voronoi tessellation $\tau_X^{(i)}$ induced by the (Cox) point process $X_H^{(i)}$ of higher–level points with parameter pair $(\gamma^{(i)}, \lambda_{H_\ell}^{(i)})$ (cmp. Section 2.4.6). These scaling properties induce that

$$\mathbb{E}_{X_H^{(1)}} \int_{L^{(1)}(\Xi_\tau^{*(1)})} \nu_1(P^{(1)}(u,o))\,du = a^2\,\mathbb{E}_{X_H^{(2)}} \int_{L^{(2)}(\Xi_\tau^{*(2)})} \nu_1(P^{(2)}(u,o))\,du\,, \tag{4.34}$$

where $L^{(i)}(\Xi_\tau^{*(i)})$ is the (typical) line system within $\Xi_\tau^{*(i)}$ and $\nu_1(P^{(i)}(u,o))$ is the length of the corresponding shortest path from u to o. Hence, by using (4.1) and (4.10), we obtain that

$$
\begin{aligned}
\gamma^{(1)} c_{sp}(\gamma^{(1)}, \lambda_{H_\ell}^{(1)}) &= \gamma^{(1)} \lambda_{H_\ell}^{(1)} \mathbb{E}_{X_H^{(1)}} \int_{L^{(1)}(\Xi_\tau^{*(1)})} c(P^{(1)}(u,o))\,du \\
&= \frac{\gamma^{(2)} \lambda_{H_\ell}^{(2)}}{a^2} \mathbb{E}_{X_H^{(1)}} \int_{L^{(1)}(\Xi_\tau^{*(1)})} c(P^{(1)}(u,o))\,du \\
&= \gamma^{(2)} \lambda_{H_\ell}^{(2)} \mathbb{E}_{X_H^{(2)}} \int_{L^{(2)}(\Xi_\tau^{*(2)})} c(P^{(2)}(u,o))\,du \\
&= \gamma^{(2)} c_{sp}(\gamma^{(2)}, \lambda_{H_\ell}^{(2)})\,,
\end{aligned}
$$

where, in the third equality, (4.34) has been used together with the assumption that the cost value of any segment of the path $P^{(i)}(u,o)$ is its length. □

4.2.8 Results of Monte–Carlo Simulations

In this section we present estimations for the two regarded characteristics of two–level hierarchical models, the mean shortest path length c_{sp} and the mean subscriber line length c_{sl}, respectively. The results of Section 4.2.8 have been obtained in cooperation with M. Rösch and are also partially documented in [88]. By using the estimators

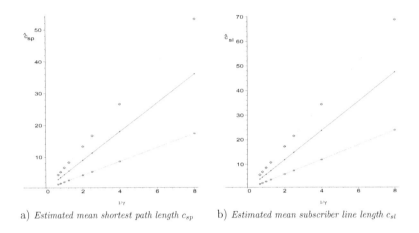

a) *Estimated mean shortest path length c_{sp}*   b) *Estimated mean subscriber line length c_{sl}*

Figure 4.10: Network characteristics for $\kappa = 10$ (o), $\kappa = 50$ (+) and $\kappa = 120$ ($\diamond$)

described in Sections 4.2.5 and 4.2.6 it is no longer necessary to simulate any lower–level points in order to estimate c_{sp} and c_{sl}. However, the most limiting factor with respect to fast runtimes remains the computation of shortest path lengths $c_A(S) = c(P(A(S), o))$ and $c_B(S) = c(P(B(S), o))$ by Dijkstra's algorithm. This computation still has to be performed for a certain set of vertices for the underlying line system as well as for their intersection points with the boundary of the typical cell Ξ_r^*. The fastest implementation of Dijkstra's algorithm is of order $O(n + m \log m)$ (cmp. [42]), where m is the size of the vertices (nodes) in the corresponding graph and n is the size of the edges. Hence, it can easily be seen that the possibility to omit a simulation of lower–level points is a huge advantage with regard to efficiency, since the simulation itself is no longer necessary and even more important the application of Dijkstra's algorithm is accelerated due to the reduced graph size with respect to the number of vertices and edges that are contained in the graph, in particular for large values of λ_{L_ℓ}. This reduced graph size is due to the fact that the lower level elements are eliminated from the graph. Especially for a large parameter $\kappa = \gamma / \lambda_{H_\ell}$, however, runtime can still become very long. In such a case we have that the size of the set of vertices of the graph considered above is quite large. Hence, we must provide an upper bound for κ with respect to evaluation. On the other hand, if κ is too small, there are few lines with relatively many higher–level points on them. For many applications, e.g. in telecommunication, this does not seem to be a very realistic assumption. Therefore, the parameter κ is investigated for $\kappa \in [10, 120]$. As a number of iterations for the estimation of the mean shortest path length as well as the mean subscriber line length $n = 50{,}000$ is used. In Figure 4.10 the scaling invariance

Table 4.2: Estimates of mean shortest path length c_{sp} and mean subscriber line length c_{sl} (without last meter) for different values of γ

a) $\kappa = 10$ b) $\kappa = 50$ c) $\kappa = 120$

γ	$\widehat{c}_{sp}$	$\widehat{c}_{sl}$
0.125	17.355	23.894
0.25	8.615	11.870
0.4	5.409	7.435
0.5	4.323	5.950
0.8	2.711	3.726
1.0	2.169	2.981
1.25	1.727	2.374
1.5	1.440	1.974

γ	$\widehat{c}_{sp}$	$\widehat{c}_{sl}$
0.125	36.074	47.494
0.25	17.972	23.665
0.4	11.267	14.815
0.5	9.003	11.857
0.8	5.618	7.397
1.0	4.499	5.920
1.25	3.600	4.735
1.5	2.996	3.942

γ	$\widehat{c}_{sp}$	$\widehat{c}_{sl}$
0.125	53.355	68.841
0.25	26.641	34.360
0.4	16.669	21.502
0.5	13.317	17.191
0.8	8.310	10.723
1.0	6.668	8.609
1.25	5.316	6.865
1.5	4.427	5.710

Table 4.3: Quotient of estimated mean shortest path length $\widehat{c}_{sp}$ and estimated mean subscriber line length $\widehat{c}_{sl}$ for different values of κ

κ	10	20	30	40	50	60	90	120
$\widehat{c}_{sp}/\widehat{c}_{sl}$	0.727	0.739	0.751	0.756	0.760	0.763	0.770	0.775

effect described in Section 4.2.7 is clearly visible. If we take the scaling parameter κ to be fixed for different values of γ then, due to (4.32) and (4.33), respectively, the estimated results for c_{sp} as well as for c_{sl} are (apart from some randomness caused by the estimator) proportional to $1/\gamma$. Hence, we see that the graphs displayed in Figure 4.10 for the different parameter values $\kappa = 10$, $\kappa = 50$, and $\kappa = 120$ are almost linear and should (at least theoretically) pass through the origin. Unfortunately, it is not possible to check the last property since this would mean that $\gamma \to \infty$. The corresponding estimated values for c_{sp} and c_{sl} are displayed in Table 4.2. Note that for the same parameter pair (γ, λ_1) we can make the observation that $\widehat{c}_{sp} < \widehat{c}_{sl}$ is astonishing but is of course favored by the fact that the last meter is chosen to be zero.

If the scaling parameter κ increases, the quotient $\widehat{c}_{sp}/\widehat{c}_{sl}$ also slightly increases. This means that the mean shortest path length c_{sp} becomes larger in relation to the mean subscriber line length c_{sl} (Table 4.3). For increasing κ we can remark that both characteristics c_{sp} as well as c_{sl} seem to increase. Obviously, a property that is at least partially responsible for this effect is that the expected area $\mathbb{E}\nu_2(\Xi_\tau^*)$ of the typical cell Ξ_τ^* of the Voronoi tessellation induced by X_H, the Cox point process of the higher-level

Table 4.4: Estimates of the mean shortest path length c_{sp} and the mean subscriber line length c_{sl} scaled by $\sqrt{\mathbb{E}(\nu_2(\Xi_\tau^*))}$

κ	10	20	30	40	50	60	90	120
$\widehat{c}_{sp}/\sqrt{\mathbb{E}(\nu_2(\Xi_\tau^*))}$	0.686	0.667	0.653	0.644	0.638	0.632	0.617	0.609
$\widehat{c}_{sl}/\sqrt{\mathbb{E}(\nu_2(\Xi_\tau^*))}$	0.944	0.903	0.870	0.852	0.840	0.828	0.801	0.786

points, also increases. In order to eliminate the effect of increasing mean areas of the typical cell it is worth looking at Table 4.4, where the characteristics c_{sp} and c_{sl} are scaled by the square root of the expected area for the typical cell, i.e., by $(\gamma\lambda_{H_\ell})^{-1/2}$. Here, we can see that for an increasing scaling parameter κ the quotient $c_{sp}/(\nu_2(\Xi_\tau^*))^{1/2}$ is decreasing for the mean shortest path length as well as $c_{sl}/(\nu_2(\Xi_\tau^*))^{1/2}$ is decreasing for the mean subscriber line length. A possible way of explaining these effects is that for increasing κ and a fixed expected area of the typical cell, the number of lines in the typical cell is increasing. Therefore, due to a higher connectivity in the cell, the mean shortest path lengths as well as the mean subscriber line lengths have a decreased value. Another possible explanation might be, as shown in Section 3.2.3, that the mean perimeter of the typical Voronoi cell induced by X_H decreases for increasing κ under the condition that $\mathbb{E}(\nu_2(\Xi_\tau^*))$, the mean area of the typical cell, is kept constant. The decreasing perimeter causes the typical cells to be more compact on average and hence values for the regarded functionals c_{sp} and c_{sl} tend to be smaller.

Recall that by Theorem 4.6 we get the following representations for c_{sp} and c_{sl}, respectively.

$$c_{sp}(\gamma, \lambda_1) = m(\kappa)\,\gamma^{-1} \qquad (4.35)$$

and

$$c_{sl}(\gamma, \lambda_1) = m'(\kappa)\,\gamma^{-1}, \qquad (4.36)$$

where $m(\kappa)$ and $m'(\kappa)$ are constants that depend only on the scaling parameter $\kappa = \gamma/\lambda_1$. By using the relationships (4.35) and (4.36), and by looking at the graphs displayed in Figure 4.10, we are able to obtain estimates $\widehat{m}(\kappa)$ and $\widehat{m}'(\kappa)$ for the slopes $m(\kappa)$ and $m'(\kappa)$ of the lines for κ being constant and $1/\gamma$ variable. Hence, due to the knowledge of $\widehat{m}(\kappa)$ and $\widehat{m}'(\kappa)$ we have the possibility to estimate c_{sp} and c_{sl} without a necessity to do simulations for any given pair of parameters $(\gamma, \lambda_{H_\ell})$, since only these parameter values need to be put into (4.35) and (4.36) in order to obtain estimates for c_{sp} and c_{sl}. From a computational point of view, we can estimate these slopes for certain discrete values of κ and afterwards a function can be fitted to the measurement points. In Figure 4.11 some values of estimated slopes as well as a fitted function are

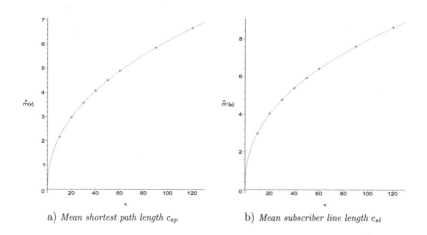

a) *Mean shortest path length c_{sp}*   b) *Mean subscriber line length c_{sl}*

Figure 4.11: Estimates for the slopes $m(\kappa)$ and $m'(\kappa)$ for different κ and the fitted function

displayed. For the fitted function we used a power function of the type

$$m(\kappa) = a\kappa^b \qquad \text{and} \qquad m'(\kappa) = a'\kappa^{b'},$$

where $a, a' \in \mathbb{R}$ and $b, b' \in (0, 1]$. Applying the least squares estimation method we obtained $a = 0.7739$, $b = 0.450$ and $a' = 1.1242$, $b' = 0.425$.

4.3 Average Distances to Nuclei in Modulated Poisson–Voronoi Tessellations

In this section another cost functional is introduced and investigated, namely the average cost from a randomly placed point to the nearest nucleus of the Voronoi tessellation τ that is induced by a modulated Poisson point process X with random driving measure Λ_X given in (2.23). The random placement might take place purely random, in other words following the distribution of a stationary Poisson point process, or again might be a modulated Poisson point process connected to the identical Boolean model Ψ that appears in the definition of the random driving measure Λ_X. The results of this section are partially based on results obtained in [28].

4.3.1 Definition of a Cost Functional

In the following let $X_H = \{X_n\}_{n \geq 1}$ be a modulated Poisson point process connected to a Boolean model Ψ with random driving measure

$$\Lambda_H(dx) = \begin{cases} \lambda_{H_1} dx & \text{if } x \in \Psi, \\ \lambda_{H_2} dx & \text{if } x \notin \Psi, \end{cases} \tag{4.37}$$

where $\lambda_{H_1}, \lambda_{H_2} \geq 0$ and $\max\{\lambda_{H_1}, \lambda_{H_2}\} > 0$. Let $\widetilde{X}_L = \{\widetilde{X}_n\}_{n \geq 1}$ be another modulated Poisson point process connected to the same Boolean model Ψ with random driving measure

$$\Lambda_L(dx) = \begin{cases} \lambda_{L_1} dx & \text{if } x \in \Psi, \\ \lambda_{L_2} dx & \text{if } x \notin \Psi, \end{cases} \tag{4.38}$$

where $\lambda_{L_1}, \lambda_{L_2} \geq 0$ and $\max\{\lambda_{L_1}, \lambda_{L_2}\} > 0$. Note that both random driving measures Λ_H and Λ_L are connected to a common Boolean model Ψ and that, given Ψ, X_H and $\widetilde{X}_L$ are assumed to be independent. Furthermore, if $N(\widetilde{X}_n)$ denotes the location of the nearest (in the Euclidean sense) point of X_H with respect to $\widetilde{X}_n \in \widetilde{X}_L$ consider the marked point process $X_L = \{\widetilde{X}_n, |\widetilde{X}_n - N(\widetilde{X}_n)|\}_{n \geq 1}$. Recall that, due to Lemma 2.6, the intensities of X_H and X_L are given by $p_\Psi \lambda_{H_1} + (1 - p_\Psi)\lambda_{H_2}$ and $p_\Psi \lambda_{L_1} + (1 - p_\Psi)\lambda_{L_2}$, respectively, where p_Ψ is the coverage probability of the Boolean model Ψ given in (2.22).

The functional we are especially interested in is the average distance $\bar{c}$ from the typical point of X_L to its nearest point of X_H. Using the Palm probability measure $P_{X_L}^*$ of X_L (cmp. (2.20)) and due to the stationarity of X_L given in Lemma 2.6 we are able to express $\bar{c}$ by

$$\bar{c} = \mathbb{E}_{X_L}|N(o)|, \tag{4.39}$$

where $\mathbb{E}_{X_L}$ is the expectation with respect to the Palm probability measure $P_{X_L}^*$. Note that due to the ergodicity of X_L (cmp. Lemma 2.7) it is possible to alternatively express the expectation $\bar{c}$ as the limit of spatial averages with respect to an averaging sequence $\{W_i\}_{i \geq 1}$ of unboundedly increasing sampling windows W_i. This means that if $\bar{c}(W)$ given by

$$\bar{c}(W) = \frac{1}{\#\{n : \widetilde{X}_n \in W\}} \sum_{n \geq 1} \mathbb{I}_W(\widetilde{X}_n)|\widetilde{X}_n - N(\widetilde{X}_n)| \tag{4.40}$$

denotes the average distance from a point of X_L to its nearest point of X_H with respect to a sampling window W and if $\{W_i\}_{i \geq 1}$ is an averaging sequence of unboundedly increasing sampling windows as defined in Section 2.1, the following holds with probability 1 (cmp. [22] and [93]).

$$\bar{c} = \lim_{i \to \infty} \bar{c}(W_i). \tag{4.41}$$

Note that the special cases of either X_H or $\widetilde{X}_L$ being stationary Poisson point processes are included in the setting described above by putting $\lambda_{H_1} = \lambda_{H_2}$ or $\lambda_{L_1} = \lambda_{L_2}$, respectively. Furthermore one can remark that an estimation that is based on $\bar{c}(W)$ might be hard to achieve due to occurring edge effects.

4.3.2 Application of Neveu's Formula

The results obtained in this section allow for a practically more feasible representation of the cost functional $\bar{c} = \mathbb{E}_{X_L}(|N(o)|)$ introduced in (4.39). Thereby a more efficient way of computing an approximation for $\bar{c}$ is derived.

Theorem 4.7 *Consider the modulated Poisson point process $X_H = \{X_n\}_{n \geq 1}$ and the (marked) modulated Poisson point process $X_L = \{\widetilde{X}_n, |\widetilde{X}_n - N(\widetilde{X}_n)|\}_{n \geq 1}$ whose random driving measures Λ_H and Λ_L are given by (4.37) and (4.38), respectively, both with respect to a common Boolean model Ψ. Let p_Ψ represent the coverage probability of Ψ defined in (2.22) and let $\lambda_H = p_\Psi \lambda_{H_1} + (1 - p_\Psi)\lambda_{H_2}$ and $\lambda_L = p_\Psi \lambda_{L_1} + (1 - p_\Psi)\lambda_{L_2}$ be the intensities of X_H and X_L, respectively. Then*

$$\bar{c} = \mathbb{E}_{X_L}(|N(o)|) = \frac{\lambda_H}{\lambda_L} \mathbb{E}_{X_H}\left(\lambda_{L_1} \int_{\Xi_\tau^* \cap \Psi} |u|du + \lambda_{L_2} \int_{\Xi_\tau^* \cap \Psi^c} |u|du\right), \qquad (4.42)$$

where Ξ_τ^* denotes the cell of the Voronoi tessellation induced by X_H which contains the origin, and where $\mathbb{E}_{X_H}$ is the expectation with respect to the Palm probability measure $P_{X_H}^*$.

Proof The proof of Theorem 4.7 is based on Neveu's exchange formula (see (2.21)) for jointly stationary point processes, which are defined on a common probability space $(\Omega, \mathcal{A}, \mathbb{P})$ equipped with some flow $\{\theta_x, \ x \in \mathbb{R}^2\}$. We use (2.21) with X_D and $\widetilde{X}_{\widetilde{D}}$ being equal to X_H and X_L, respectively. Thus, the mark space $\mathbb{M}$ will be omitted and $\widetilde{\mathbb{M}} = [0, \infty)$. Consider the function $f : \mathbb{R}^2 \times [0, \infty) \times \Omega \to [0, \infty)$ given by

$$f(x, \widetilde{g}, \omega) = \begin{cases} \widetilde{g} & \text{if } X_H(\theta_{-x}\omega, B_{|x|}^{\neq}(x)) = 0, \\ 0 & \text{otherwise} \end{cases} \qquad (4.43)$$

for any $x \in \mathbb{R}^2$, $\widetilde{g} \geq 0$, and $\omega \in \Omega$, where $B_{|x|}^{\neq}(x) = \{y \in \mathbb{R}^2 : |y - x| < |x|\}$. Note that if $x \in \mathbb{R}^2$ is an atom of the counting measure $X_H(\omega, \cdot)$, then $f(-x, \widetilde{g}, \omega) = \widetilde{g}$ only if there are no other atoms of $X_H(\omega, \cdot)$ which have a distance of less than $|x|$ to the origin. Hence by applying Neveu's exchange formula (2.4) we obtain that

$$\begin{aligned} \mathbb{E}_{X_L}(|N(o)|) &= \int_{\Omega \times \widetilde{\mathbb{M}}} \int_{\mathbb{R}^2} f(-x, \widetilde{g}, \omega) X_H(\omega, dx) P_{X_L}^*(d(\omega, \widetilde{g})) \\ &= \frac{\lambda_H}{\lambda_L} \int_\Omega \int_{\mathbb{R}^2 \times \widetilde{\mathbb{M}}} f(x, \widetilde{g}, \theta_x\omega) X_L(\omega, d(x, \widetilde{g})) P_{X_H}^*(d\omega). \end{aligned} \qquad (4.44)$$

Given the Boolean model Ψ the inner integral on the right–hand side of (4.44) can be expressed as

$$\int_{\mathbb{R}^2 \times \tilde{\mathbb{M}}} f(x, \tilde{g}, \theta_x \omega) X_L(\omega, d(x, \tilde{g})) = \int_{(\mathbb{R}^2 \cap \Psi) \times \tilde{\mathbb{M}}} f(x, \tilde{g}, \theta_x \omega) X_L(\omega, d(x, \tilde{g}))$$

$$+ \int_{(\mathbb{R}^2 \cap \Psi^c) \times \tilde{\mathbb{M}}} f(x, \tilde{g}, \theta_x \omega) X_L(\omega, d(x, \tilde{g})). \tag{4.45}$$

Additionally, given Ψ and the Voronoi cell Ξ^*_τ of X_H that contains the origin, note that the random number of points of X_L in $\Xi^*_\tau \cap \Psi$ is Poisson distributed with expectation $\eta_1 = \lambda_{L_1} |\Xi^*_\tau \cap \Psi|$, while the random number of points of X_L in $\Xi^*_\tau \cap \Psi^c$ is Poisson distributed with expectation $\eta_2 = \lambda_{L_2} |\Xi^*_\tau \cap \Psi^c|$. Hence, by the definition of the function f given in (4.43) the first integral on the right side of (4.45) can be written as

$$\int_{(\mathbb{R}^2 \cap \Psi) \times \tilde{\mathbb{M}}} f(x, \tilde{g}, \theta_x \omega) X_L(\omega, d(x, \tilde{g})) = \sum_{k=1}^{\infty} e^{-\eta_1} \frac{\eta_1^k}{k!} \int_{\Xi^*_\tau \cap \Psi} \cdots \int_{\Xi^*_\tau \cap \Psi} \sum_{i=1}^{k} \frac{|u_i|}{|\Xi^*_\tau \cap \Psi|^k} du_1 \ldots du_k \,,$$

due to the independence and the conditional uniform distribution of the points of X_L in $\Xi^* \cap \Psi$. Therefore,

$$\int_{(\mathbb{R}^2 \cap \Psi) \times \tilde{\mathbb{M}}} f(x, \tilde{g}, \theta_x \omega) X_L(\omega, d(x, \tilde{g})) = \sum_{k=1}^{\infty} e^{-\eta_1} \frac{\eta_1^k}{k!} \frac{k}{|\Xi^*_\tau \cap \Psi|} \int_{\Xi^*_\tau \cap \Psi} |u| \, du$$

$$= \lambda_{L_1} \int_{\Xi^*_\tau \cap \Psi} |u| \, du \,.$$

Analogously, it can be shown that

$$\int_{(\mathbb{R}^2 \cap \Psi^c) \times \tilde{\mathbb{M}}} f(x, \tilde{g}, \theta_x \omega) X_L(\omega, d(x, \tilde{g})) = \sum_{k=1}^{\infty} e^{-\eta_2} \frac{\eta_2^k}{k!} \frac{k}{|\Xi^*_\tau \cap \Psi^c|} \int_{\Xi^*_\tau \cap \Psi^c} |u| \, du$$

$$= \lambda_{L_2} \int_{\Xi^*_\tau \cap \Psi^c} |u| \, du \,.$$

Altogether we get that

$$\mathbb{E}_{X_L}(|N(o)|) = \frac{\lambda_H}{\lambda_L} \mathbb{E}_{X_H} \left(\int_{\Xi^*_\tau \cap \Psi} \lambda_{L_1} |u| du + \int_{\Xi^*_\tau \cap \Psi^c} \lambda_{L_2} |u| du \right),$$

which completes the proof of the theorem.   □

In the special case that $\lambda_{L_1} = \lambda_{L_2}$, i.e., $\{\tilde{X}_L\}$ is a stationary Poisson point process Theorem 4.7 can be restated as follows.

Corollary 4.3 *Supppose that $\lambda_{L_1} = \lambda_{L_2}$, i.e., $\{\widetilde{X}_L\}$ is a stationary Poisson point process with intensity λ_L. Then*

$$\bar{c} = \mathbb{E}_{X_L}(|N(o)|) = \lambda_H \mathbb{E}_{X_H} \int_{\Xi_\tau^*} |u| du. \qquad (4.46)$$

Note that (4.46) induces in particular that $\bar{c}$ is independent of λ_L. In the Poisson case, i.e., if $\lambda_{H_1} = \lambda_{H_2}$ and $\lambda_{L_1} = \lambda_{L_2}$, the cost functional $\bar{c} = \mathbb{E}_{X_L}|N(o)|$ can be analytically computed as (see also [4] and [16])

$$\bar{c} = \lambda_H \mathbb{E}_{X_H} \int_{\Xi^*} |u| du = \lambda_H \int_{\mathbf{R}^2} |u| \exp\left(-\lambda_H \pi |u|^2\right) du = \frac{1}{2\sqrt{\lambda_H}}. \qquad (4.47)$$

4.3.3   Estimation Based on Monte-Carlo Simulation

Theorem 4.7 induces a useful approach for the construction of an estimator for the cost functional $\bar{c}$ which is given in the following lemma.

Lemma 4.4 *Let $\{(\Xi_1^*, \Psi_1^*)..., (\Xi_n^*, \Psi_n^*)\}$ be a sequence of independent copies of (Ξ_τ^*, Ψ) under the Palm probability measure $P_{X_H}^*$. The estimator $\widehat{\bar{c}}$ given by*

$$\widehat{\bar{c}} = \frac{\lambda_H}{\lambda_L} \frac{1}{n} \sum_{i=1}^{n} \int_{\Xi_{\tau,i}^*} |u| \widetilde{\Lambda}_{L,i}(du), \qquad (4.48)$$

where

$$\widetilde{\Lambda}_{L,i}(dx) = \begin{cases} \lambda_{L_1} dx & \text{if } x \in \Psi_i^*, \\ \lambda_{L_2} dx & \text{if } x \notin \Psi_i^*, \end{cases} \qquad (4.49)$$

is an unbiased and consistent estimator for $\bar{c}$.

The estimator $\widehat{\bar{c}}$ given in Lemma 4.4 will be used in Section 4.3.4 in order to obtain numerical results for some sample scenarios. Note that if $\lambda_{L_1} = \lambda_{L_2}$ then the integral $\int_{\Xi_{\tau,i}^*} |u| \widetilde{\Lambda}_{L,i}(du)$ is computed analytically, otherwise it is computed via numerical approximation. This is due to the fact that in the first case, by applying (4.46), we are able to rewrite the integral as an integral with respect to the Lebesgue measure. If $\lambda_{L_1} \neq \lambda_{L_2}$ integration must be performed with respect to the measure $\widetilde{\Lambda}_{L,i}$ and therefore the shape of Ψ_i^* has to be taken into account. Another important fact concerning a numerical evaluation is that it is not necessary to simulate any points of X_L in order to apply the estimator $\widehat{\bar{c}}$ given in (4.48).

4.3.4   Numerical Examples

The results of Section 4.3.4 have been obtained in cooperation with K. Posch and A. Upowsky and are partially documented in their respective diploma theses ([84], [107]). With respect to the parameters β, the intensity of the germs of the Boolean model Ψ, p_Ψ, the coverage probability of Ψ introduced in (2.22) and r, the fixed radius of the grains of Ψ we regard the same values as in the first example of Section 3.3.3, i.e., we have that $\beta = 0.2$, $p_\Psi = 0.6$ and therefore, due to (2.22), it is obtained that $r = 1.20761$. With regard to the estimated values of the cost functional $\bar{c}$ defined in Section 4.3.2 it can be stated that it increases as λ_{H_1} tends to 0 (Figure 4.12a) with respect to a fixed intensity $\lambda_H = 12$. Note that for this example the intensities of the process X_L are assumed to be equal, i.e., $\lambda_{L_1} = \lambda_{L_2}$ The sample size is $n = 2{,}000{,}000$ for each pair of parameters $(\lambda_{H_1}, \lambda_{H_2})$. The effect of an increasing value of the cost functional if λ_{H_1} tends to 0 can possibly be explained by the appearance of cells that have a relatively large ratio of perimeter to area that causes a relatively large mean distance to the cell nuclei. Note that in the case of a Poisson–Voronoi tessellation ($\lambda_{H_1} = \lambda_{H_2}$) the estimated value for the mean distance to the cell nuclei of 0.14437 coincides well with the theoretical value of $(2\sqrt{\lambda_H})^{-1} = 0.14434$.

As a second numerical example we have a look at a scenario where $\lambda_{L_1} \neq \lambda_{L_2}$. For this scenario we take the same values for β, p_Ψ and r as above and additionally keep $\lambda_{H_1} = 4$ and $\lambda_{H_2} = 24$ fixed. The values for λ_{L_1} and λ_{L_2} are varied under the condition that $\lambda_L = 1$. The results shown in Figure 4.12b reflect quite well the linear relationship between the value of λ_{L_1} and the estimated cost functional $\hat{\bar{c}}$ in this case. This linear relationship is a direct consequence of (4.42). Due to this linear relationship it is sufficient to estimate two expectations $\mathbb{E}_X(\int_{\Xi_X \cap \Psi} |u| du)$ and $\mathbb{E}_X(\int_{\Xi_X \cap \Psi^c} |u| du)$ for a specific pair of parameters λ_{L_1} and λ_{L_2} in order to obtain estimates of $\bar{c}$ for all pairs of parameters λ_{L_1} and λ_{L_2} based on (4.42).

Notice that numerical evaluations of examples where $\lambda_{L_1} \neq \lambda_{L_2}$ are more time consuming due to the numerical computation of the estimator $\hat{\bar{c}}$ introduced in (4.48), as opposed to the case where $\lambda_{L_1} = \lambda_{L_2}$ since here $\hat{\bar{c}}$ is computed analytically, given realisations of the typical cell (cmp. Section 4.3.2). Therefore, for cases where $\hat{\bar{c}}$ had to be computed numerically we took $n = 100{,}000$.

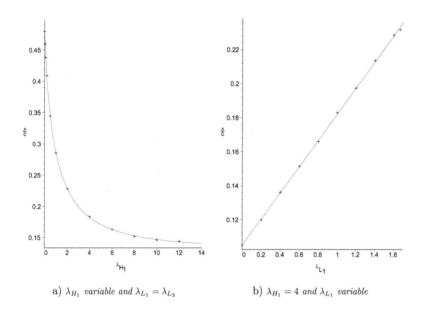

a) λ_{H_1} variable and $\lambda_{L_1} = \lambda_{L_2}$ b) $\lambda_{H_1} = 4$ and λ_{L_1} variable

Figure 4.12: Estimated cost functionals for $\lambda_H = 12$ fixed

Chapter 5

Testing Methods for Programs with Random Input or Output

In Chapters 3 and 4 some algorithms for simulations of the typical cell for different random tessellation models as well as for the estimation of related cost functionals with respect to two–level hierarchical models have been introduced. A major problem that is connected with actual simulations and computations is how to ensure that the implementations based on these algorithms deliver results that are correct, this means in our context provide correct results for the characteristics to be evaluated. This problem leads to a necessity for the application of software tests, in particular of software tests for software that has a random input or output. In the following chapter some fundamental principles of such software tests for software with random input or output are introduced. These fundamental principles are then applied in order to develop tests for implementations of the algorithms described in Chapters 3 and 4. Three approaches for testing the implementations are considered in particular. In Section 5.2 tests based on a statistical oracle are introduced and applied. A statistical oracle that is deduced from some known theoretical relationships is used in order to get inference about the correctness of the implementation. In Section 5.3 the statistical oracle is combined with a technique called metamorphic testing. Metamorphic testing means a simultaneous testing of different test cases that are connected via a metamorphic relation, where a metamorphic relation represents an expected relation among the related inputs and outputs of multiple executions for the implementation unit under test (IUT). An example for such a combined testing method of a statistical oracle and a metamorphic relation is provided. Finally, in Section 5.4, the techniques derived in Sections 5.2 and 5.3 are used in order to develop software tests that are based on the comparison to an already tested (and validated) gold–standard implementation. A slight modification of this testing technique is to test two implementations versus each other. For more detailed information on software tests in general and on some of the techniques applied in this thesis the reader is, for example, referred to [14], [20], [39], [63], and [64].

5.1 General Principles of Software Testing

A *software test* is a possible method in computer science for the partial verification and validation of an IUT. It is the process used to identify the correctness, completeness, security and quality of developed computer software. In particular, software testing means to compare the actual behaviour of the IUT with the specifications or expected behaviour of the software by the help of constructed test cases. In this context a *test case* is a set of conditions or variables (in our case often input parameters) under which a tester can determine if a requirement of an application is fully or at least partially satisfied. Usually, a test case possesses a known input and an expected output which is determined before the actual test is executed. Hence, the known input should test a precondition, whereas the expected output tests a postcondition.

In the following we will focus ourselves on black–box testing, where black–box means that, contrary to white–box testing, the IUT considered is only accessed through the same interfaces that a customer or user would use. More specifically we will focus on function–oriented tests where the main aim of the test cases is to check for functional correctness of the IUT in the sense that, given a set of input parameters, correct values for the output parameters are computed. The principal structure of the software tests considered in this thesis is described in Figure 5.1. The input parameters, in our cases mostly values for the parameters of the mathematical model, are provided to the IUT via a suitable interface. The IUT then computes the actual results that are afterwards, once more via a suitable interface, given to the analyser. The analyser transforms the actual results in characteristics that the comparator can use to compare them to the expected output. Thereby it can be determined if the test passes or not. The problem of how to obtain a useful analyser and a subsequent comparator is also known as the test oracle problem. A possible solution with respect to software that has a random output will be discussed in the following.

5.2 Testing Based on Statistical Oracles

5.2.1 Basic Principle

A main aspect of software testing is the oracle problem. This means, as described in Section 5.1, that rules or conditions must be provided that are able to indicate whether

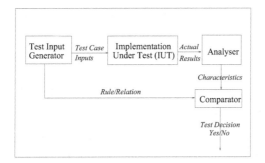

Figure 5.1: Structure of a software test

a test passes or not. In other words a construction principle for the comparator of a software test has to be found. Then, a method has to be given that can be used by the analyser in order to transform the actual results of the IUT, given the test input parameters, into characteristics that can be used by the comparator to decide whether the test passes or not.

In the following we will introduce the statistical oracle that can be considered as a special case of the Heuristic Oracle ([39]) and the Parametric Oracle ([14]), respectively. The basis for a statistical oracle is given by some statistical methods, especially statistical tests where the intent is to verify some statistical characteristics like the first–order or higher–order moments of the actual test results. In practice, this means that the analyser for our software test is given by a statistical analyser that for example consists of an estimator for moments of a specific characteristic based on a sample provided by the IUT. These estimated moments, which are of a random nature, are afterwards given to the comparator. In the case of a statistical oracle we have that the comparator itself is also of a random nature, meaning that the comparison between the characteristics provided by the statistical analyser to the expected output are performed by the help of statistical test methods. Hence, due to the random nature of the analyser as well as the comparator, we can not expect that the decision of a statistical oracle is always correct in a deterministic sense but only in the sense of the statistical tests involved in the comparator. An advantage, on the other hand, is that statistical oracles are useable for software with random inputs or outputs and for randomly generated test cases.

Note that all tests considered in this chapter are at least partially based on a statistical oracle. In the examples of this section we will consider tests that are almost purely based on a statistical oracle while in the following sections tests are constructed that utilise a statistical oracle in conjunction with metamorphic testing relationships and with results from gold–standard implementations or with results from a second implementation, respectively.

5.2.2 Examples

As an example for a test of software with random output we have a look at the simulation of the typical cell for different types of random tessellations that is described in Sections 3.1.3 and 3.3. Given a random tessellation τ that is induced by a (stationary) random point process X we are able to use (2.25) in order to construct a statistical oracle. Recall that (2.25) states that

$$\mathbb{E}\left[\nu_2(\Xi_\tau^*)\right] = \frac{1}{\lambda_\tau},$$

where τ is a random tessellation with finite and positive intensity λ_τ and where Ξ_τ^* is the typical cell of τ. Note that for the cases of Voronoi tessellations that will be regarded in the following we have that $\lambda_\tau = \lambda_X$, where λ_X represents the intensity of the random point process X that induces the Voronoi tessellation τ. In other words the fact that the mean area of the typical cell for a tessellation τ is given by the reciprocal of the corresponding intensity λ_X of X induces the following null–hypothesis under test. The expected results for the area of the typical cell Ξ_τ^* provided by the implemented algorithm should be equal to λ_X^{-1} for various intensities λ_X. To evaluate such a hypothesis a well–known statistical test can be used. Let $\tilde{\xi}_{\tau,1}^*, .., \tilde{\xi}_{\tau,n}^*$ be n realisations of the implemented version $\tilde{\Xi}_\tau^*$ of the typical cell Ξ_τ^*, where n is supposed to be large. These realisations are generated in order to get an estimate $\frac{1}{n}\sum_{i=1}^{n}\tilde{\nu}_2(\tilde{\xi}_{\tau,i}^*)$ for $\mathbb{E}\nu_2(\Xi_\tau^*)$, where $\tilde{\nu}_2(\tilde{\xi}_{\tau,i}^*)$ is the result for the area of a particular realisation $\tilde{\xi}_{\tau,i}^*$ provided by the IUT. Due to the fact that we suppose the underlying variances of $\tilde{\nu}_2(\tilde{\Xi}_{\tau,i}^*)$ to be finite and the different realisations of $\tilde{\Xi}_\tau^*$ to be independent and identically distributed and since the sample size n is supposed to be large, the test statistic

$$T = \sqrt{n}\,\frac{\frac{1}{n}\sum_{i=1}^{n}\tilde{\nu}_2(\tilde{\Xi}_{\tau,i}^*) - (\lambda_X)^{-1}}{\sqrt{\frac{1}{n-1}\sum_{i=1}^{n}\left(\tilde{\nu}_2(\tilde{\Xi}_{\tau,i}^*) - \frac{1}{n}\sum_{j=1}^{n}\tilde{\nu}_2(\tilde{\Xi}_{\tau,j}^*)\right)^2}} \tag{5.1}$$

is approximately standard normal distributed due to the Central Limit Theorem given in Theorem A.1. Hence, an asymptotic Gaussian test can be applied in this situation in order to obtain inference about the null–hypothesis and thereby in order to obtain inference about the correctness of the implementation. We construct a statistical analyser that, given realisations of the typical cell Ξ_τ^* computes the test statistic T. Then the actual value of T can be used by the comparator to determine whether the test passes or not by the help of an asymptotic Gaussian test for the null-hypothesis that is based on (2.25).

Tests of this type have proven to be very sensitive for the detection of errors during our experiments, especially for very large numbers of iterations n which is equivalent to a very large sample size for the compuation of the test statistic in the statistical analyser.

An alternative approach (cmp. [64]) to the one presented in this section would be to test for inequalities rather than to test for equality. For example, some fixed $\varepsilon > 0$ can be defined and hypotheses like $\mathbb{E}(\tilde{\nu}_2(\tilde{\Xi}_\tau^*)) \notin [(\lambda_X)^{-1} - \varepsilon, (\lambda_X)^{-1} + \varepsilon]$ can be tested. From a statistical point of view, this approach leads to intersection–union tests (cmp. [19], Chapter 8) and to test statistics of the form

$$T_1 = \sqrt{n} \; \frac{\frac{1}{n}\sum_{i=1}^{n} \tilde{\nu}_2(\tilde{\Xi}_{\tau,i}^*) - (\lambda_X)^{-1} + \varepsilon}{\sqrt{\frac{1}{n-1}\sum_{i=1}^{n} \left(\tilde{\nu}_2(\tilde{\Xi}_{\tau,i}^*) - \frac{1}{n}\sum_{j=1}^{n} \tilde{\nu}_2(\tilde{\Xi}_{\tau,j}^*)\right)^2}} \tag{5.2}$$

and

$$T_2 = \sqrt{n} \; \frac{\frac{1}{n}\sum_{i=1}^{n} \tilde{\nu}_2(\tilde{\Xi}_{\tau,i}^*) - (\lambda_X)^{-1} - \varepsilon}{\sqrt{\frac{1}{n-1}\sum_{i=1}^{n} \left(\tilde{\nu}_2(\tilde{\Xi}_{\tau,i}^*) - \frac{1}{n}\sum_{j=1}^{n} \tilde{\nu}_2(\tilde{\Xi}_{\tau,j}^*)\right)^2}} \tag{5.3}$$

where, under the null–hypothesis, the statistics T_1 and T_2 are assumed to be nearly normal distributed. Compared to the first approach based on equality that will be used in the following, this approach has some advantages as well as disadvantages. First of all, it is possible by using the approach based on inequalities to control the error of classifying an incorrect IUT as correct. But, apart from an increased complexity of the testing method and apart from the fact that, for example, a choice of the parameter ε is not obvious, a main disadvantage is that the probability for a misclassification of a correct IUT as incorrect can not be arbitrarily fixed. Naturally, this might lead to a danger of searching for non–existent implementation errors.

We now turn our attention to more concrete examples for software tests based on a statistical oracle that is induced by the relationship given in (2.25). First we have a look at the typical cell Ξ_τ^* of the Voronoi tessellation τ that is induced by a Cox point process X_c with random driving measure Λ_{X_c} introduced in Section 2.4.6. Recall that Lemma 2.10 states that for this type of tessellation the intensity λ_c of the generating point process X_c is given by $\lambda_c = \lambda_\ell \gamma$, where γ denotes the intensity of the underlying Poisson line process X_ℓ, while λ_ℓ is the (linear) intensity of the Cox point process on the Poisson lines. Therefore, we obtain a test statistic of the form

$$T = \sqrt{n} \; \frac{\frac{1}{n}\sum_{i=1}^{n} \tilde{\nu}_2(\tilde{\Xi}_{\tau,i}^*) - (\lambda_c)^{-1}}{\sqrt{\frac{1}{n-1}\sum_{i=1}^{n} \left(\tilde{\nu}_2(\tilde{\Xi}_{\tau,i}^*) - \frac{1}{n}\sum_{j=1}^{n} \tilde{\nu}_2(\tilde{\Xi}_{\tau,j}^*)\right)^2}} \tag{5.4}$$

Table 5.1: Area tests for the typical Voronoi cell induced by a Cox point process by means of p–values

γ	0.125	0.25	0.4	0.5	0.8	1.0	1.25	1.5
$\kappa=10$	0.994	0.608	0.972	0.675	0.958	0.979	0.582	0.158
$\kappa=50$	0.778	0.693	0.932	0.917	0.082	0.114	0.002	0.798
$\kappa=120$	0.092	0.745	0.760	0.434	0.436	0.880	0.347	0.306

which is (for large n) approximately standard normal distributed.

Table 5.1 shows p–values of such a test for different values of γ, scaling invariance parameter $\kappa = \gamma/\lambda_\ell$, introduced in Section 2.4.6, and a sample size of $n = 2,000,000$. Recall that the meaning of the p–value is that the null–hypothesis is rejected if the chosen significance level α is larger or equal to the p–value. For a (given) significance level $\alpha = 0.05$ the null–hypothesis is therefore rejected once for all regarded cases ($\gamma = 1.25$ and $\kappa = 50$). The one significant rejection among the 24 test cases coincides quite well with the definition of the significance level since, supposed the IUT is correct, we would (theoretically) expect $24 * 0.05 = 1.2$ rejection cases. Due to the fact that the number of rejections we obtained is even smaller than the expected number under the null–hypothesis we can assume that the IUT provides correct values for the mean area of the typical Voronoi cell induced by a Cox point process with random driving random measure given in (2.31).

As a second example consider the typical Voronoi cell that is induced by a modulated Poisson process as defined in Section 2.4.7. Recall that by Lemma 2.6 we have that the intensity of the generating modulated Poisson point process X is given by $\lambda_X = p_\Psi \lambda_{X_1} + (1 - p_\Psi)\lambda_{X_2}$. Hence, using once more (2.25), we obtain a test statistic to be applied that can be written as

$$T = \sqrt{n} \frac{\frac{1}{n}\sum_{i=1}^{n} \tilde{\nu}_2(\tilde{\Xi}^*_{\tau,i}) - (p_\Psi \lambda_{X_1} + (1 - p_\Psi)\lambda_{X_2})^{-1}}{\sqrt{\frac{1}{n-1}\sum_{i=1}^{n} \left(\tilde{\nu}_2(\tilde{\Xi}^*_{\tau,i}) - \frac{1}{n}\sum_{j=1}^{n} \tilde{\nu}_2(\tilde{\Xi}^*_{\tau,j})\right)^2}}, \qquad (5.5)$$

where, for large sample size n, T is nearly standard normal distributed. In Table 5.2 some results of testing in form of p–values for different input values of λ_{X_1} and $\kappa' = (p, \lambda_{X_1}/\beta, \lambda_{X_2}/\beta)$ are displayed. Recall that κ' denotes the scaling invariance vector for a Voronoi tessellation that is induced by a modulated Poisson point process (cmp. Section 2.4.7). For a test level of $\alpha = 0.05$ the null–hypothesis is rejected in a number of cases that is smaller than the expected one (1.2 cases). Therefore, it is justified to assume that the IUT provides correct estimates for the mean area of the typical cell of a Voronoi tessellation generated by a modulated Poisson process.

Table 5.2: Area tests for the typical Voronoi cell induced by a modulated Poisson process by means of p–values

λ_{X_1}	60	70	80	90	100	110	120	130
$\kappa'=(0.2;\ 100;\ 10)$	0.725	0.644	0.351	0.495	0.628	0.945	0.321	0.059
$\kappa'=(0.2;\ 100;\ 20)$	0.656	0.364	0.728	0.960	0.752	0.936	0.234	0.195
$\kappa'=(0.8;\ 50;\ 5)$	0.379	0.066	0.098	0.187	0.742	0.433	0.510	0.904

Another example for a test based on a statistical oracle is related to the computation of the cost functional $\bar{c}$ introduced in Section 4.3. In particular we have a look at the special case, where $\lambda_{H_1} = \lambda_{H_2}$ and $\lambda_{L_1} = \lambda_{L_2}$. This means that the modulated Poisson point process is indeed a Poisson point process and that the corresponding Voronoi tessellation is a Poisson–Voronoi tessellation. Hence, we obtain for $\lambda_H = \lambda_{H_1} = \lambda_{H_2}$ and for $\lambda_{L_1} = \lambda_{L_2} = 1$ that (cmp. [78])

$$\mathbb{E}_{X_H} \int_{\Xi_\tau^*} |u|du = \frac{1}{2\sqrt{\lambda_H}^3} \tag{5.6}$$

and that therefore

$$\bar{c} = \lambda_H \mathbb{E}_{X_H} \int_{\Xi_\tau^*} |u|du = \frac{1}{2\sqrt{\lambda_H}} \tag{5.7}$$

These equations induce a test statistic T given by

$$T = \sqrt{n}\ \frac{\frac{1}{n}\sum_{i=1}^n \int_{\hat{\Xi}_{\tau,i}^*} |u|du - \frac{1}{2\sqrt{\lambda_H}^3}}{\sqrt{\frac{1}{n-1}\sum_{i=1}^n \left(\int_{\hat{\Xi}_{\tau,i}^*} |u|du - \frac{1}{n}\sum_{j=1}^n \int_{\hat{\Xi}_{\tau,j}^*} |u|du\right)^2}}, \tag{5.8}$$

which for a large sample size n is approximately standard normal distributed. In Table 5.3 resulting p–values are shown for different values of λ_H and a sample size of $n = 1{,}000{,}000$. None of the regarded test cases shows a rejection for a test level of $\alpha = 0.05$. Hence, it is justified to say that the implementation of the cost functional $\bar{c}$ provides correct results at least with respect to Poisson–Voronoi tessellations.

5.3 Statistical Metamorphic Testing

5.3.1 Basic Principle

In the absence of a statistical oracle that is directly usable or as an extension to tests based on statistical oracles, a technique called metamorphic testing can be applied to

Table 5.3: Tests for accuracy of the implementation for the cost functional $\bar{c}$ with respect to Poisson–Voronoi tessellations

λ_H	10	20	30	40	50	60	70
p–value	0.8109	0.3101	0.881	0.4416	0.8589	0.8159	0.5903
λ_H	80	90	100	110	120	130	140
p–value	0.5816	0.9711	0.366	0.2849	0.6072	0.1375	0.3341
λ_H	150	160	170	180	190	200	210
p–value	0.7617	0.4691	0.2479	0.5362	0.0353	0.9433	0.2173

mathematical, in particular statistical, software. More precisely, metamorphic testing means the simultaneous test of different test cases that are connected via a *metamorphic relation* (MR), where a MR is an expected relation among the related inputs and outputs of multiple executions of the IUT. Examples for MR are numerous among mathematical problems, for example, there are various MR for matrices with real–valued entries (cmp. [63]) or for trigonometrical functions (cmp. [20]). Metamorphic testing can be considered as a kind of a self testing approach (cmp. [17] and [18]) which means that the IUT is tested vs. itself. In general, we check two related test cases t_1 and t_2 and their respective outputs $o(t_1)$ and $o(t_2)$, given by the IUT, against the MR. If the MR can not be satisfied then an error in the IUT is indicated.

As an example for a deterministic MR regard the greatest common divisor $gcd(a, b)$ of two natural numbers a and b. It is a well-known fact that the greatest common divisor is commutative in the sense that

$$gcd(a, b) = gcd(b, a). \tag{5.9}$$

Now, we can use (5.9) as a MR in the following way. Consider an IUT that should compute the greatest common divisor of two natural numbers and a test case t_1, where the input parameters are given by $a, b \in \mathbb{N}$. Then, the choice of t_1 and the MR implies that the test case t_2 should have input parameters b and a and that with respect to the outcomes $o(t_1)$ and $o(t_2)$ of t_1 and t_2, respectively, we have that $o(t_1) = o(t_2)$ if the IUT works correctly. Therefore, in this example, the analyser simply provides the two values $o(t_1)$ and $o(t_2)$ to the comparator. Afterwards the comparator checks the metamorphic relationship $o(t_1) = o(t_2)$ in order to decide whether the test passes or not.

With regard to the applications we want to have a look at in this thesis, settings are a bit different since the IUT typically produces a random output. Therefore the testing technique using metamorpic relations has to be slightly modified in order to account for this property of the IUT. More specifically, we do not use deterministic metamorphic relations as in the example for the greatest common divisor. Instead statistical metamorphic relations are applied in order to check whether the test passes or not. This means that the tests regarded in the following represent a combination of tests using metamorphic relations and a statistical oracle as described in Secion 5.2 Therefore, we obtain a comparator that is on the one hand based on a statistical oracle, but on the other hand also on a (statistical) MR which means, for example, a MR for first–order or higher–order moments of random characteristics. Of course, as in Section 5.2, we then have to regard the results of software testing not in a deterministic sense but in the sense of results for a statistical test.

5.3.2 Examples

In our examples that are based on the algorithm introduced in Section 3.3 the scaling invariance effect explained in Section 2.4.6 is used in order to derive statistical metamorphic relations. If we consider the mean perimeters $\mathbb{E}\nu_1(\partial\Xi_\tau^{*(1)})$ and $\mathbb{E}\nu_1(\partial\Xi_\tau^{*(2)})$ of the typical cell of a Cox–Voronoi tessellation induced by a Cox point process defined in Section 2.4.6 for two different pairs of parameters $(\gamma^{(1)}, \lambda_\ell^{(1)})$ and $(\gamma^{(2)}, \lambda_\ell^{(2)})$, where $\kappa = \gamma^{(1)}/\lambda_\ell^{(1)} = \gamma^{(2)}/\lambda_\ell^{(2)}$, we obtain that

$$\gamma^{(1)}\mathbb{E}\nu_1(\partial\Xi_\tau^{*(1)}) = \gamma^{(2)}\mathbb{E}\nu_1(\partial\Xi_\tau^{*(2)}). \tag{5.10}$$

The relation given in (5.10) can now be used in the sense of metamorphic testing as a statistical MR. Consider a test case t_1 given by $t_1 : (\gamma, \lambda_\ell) = (\gamma^{(1)}, \lambda_\ell^{(1)})$. Then, the choice of t_1 and (5.10) induce that the second test case t_2 should be of the form $t_2 : (\gamma, \lambda_\ell) = (\gamma^{(2)}, \lambda_\ell^{(2)})$, where $\gamma^{(1)}/\lambda_\ell^{(1)} = \gamma^{(2)}/\lambda_\ell^{(2)} = \kappa$. Note that if one of the two test cases is regarded individually no direct statistical oracle is available with respect to the mean perimeter of the typical cell. Instead, under the assumption that we have an equal sample size n for both test cases, the statistical MR given in (5.10) allows us to construct a test statistic of the form

$$T = \frac{\sum_{i=1}^n \left(\gamma^{(1)}\tilde{\nu}_1(\partial\tilde{\Xi}_{\tau,i}^{*(1)}) - \gamma^{(2)}\tilde{\nu}_1(\partial\tilde{\Xi}_{\tau,i}^{*(2)})\right)}{\sqrt{(n-1)(S_{(1)}^2 + S_{(2)}^2)}}, \tag{5.11}$$

where in this case

$$S_{(1)}^2 = \sum_{i=1}^n \left(\gamma^{(1)}\tilde{\nu}_1(\partial\partial\tilde{\Xi}_{\tau,i}^{*(1)}) - \frac{1}{n}\sum_{j=1}^n \gamma^{(1)}\tilde{\nu}_1(\partial\tilde{\Xi}_{\tau,j}^{*(1)})\right)^2$$

and

$$S_{(2)}^2 = \sum_{i=1}^n \left(\gamma^{(2)}\tilde{\nu}_1(\partial\tilde{\Xi}_{\tau,i}^{*(2)}) - \frac{1}{n}\sum_{j=1}^n \gamma^{(2)}\tilde{\nu}_1(\partial\tilde{\Xi}_{\tau,j}^{*(2)})\right)^2.$$

Table 5.4: Tests for equality of expected perimeter estimates for $\kappa = 50$ by means of p–values

$\gamma^{(1)}\backslash\gamma^{(2)}$	0.125	0.25	0.4	0.5	0.8	1.0	1.25	1.5
0.125	–	0.636	0.373	0.393	0.918	0.928	0.995	0.437
0.25	0.636	–	0.251	0.268	0.851	0.867	0.986	0.306
0.4	0.373	0.251	–	0.521	0.957	0.963	0.998	0.566
0.5	0.3933	0.268	0.521	–	0.951	0.958	0.998	0.544
0.8	0.918	0.851	0.957	0.951	–	0.527	0.878	0.061
1.0	0.928	0.86679	0.962	0.958	0.527	–	0.863	0.053
1.25	0.995	0.986	0.998	0.998	0.878	0.863	–	0.003
1.5	0.437	0.306	0.566	0.544	0.061	0.053	0.003	–

The test statistic given in (5.11) is nearly standard normal distributed for large sample sizes n of the two test cases t_1 and t_2 which allows for the construction of an asymptotic Gaussian test. Therefore, if the analyser computes the values of the test statistic T based on the ouputs $o(t_1)$ and $o(t_2)$ of t_1 and t_2, respectively, the comparator is able to decide whether the test passes or not based on the value of T provided by the analyser and on an asymptotic Gaussian test.

In Table 5.4 some results for the statistical metamorphic testing procedure are described given our IUT, sample sizes $n = 2,000,000$ for t_1 and t_2 each and a scaling invariance parameter $\kappa = 50$. The p–values that are displayed indicate that the scaling invariance effect is reflected quite well by the test results since, given 28 pairs of parameters, only for one pair $(\gamma^{(1)}, \gamma^{(2)}) = (1.25, 1.5)$ we have a p–value of less than 0.05. Hence, it is deducable that our IUT passes the test based on the metamorphic relation given in (5.10).

In order to derive a second example for a test based on the combination of metamorphic testing and a statistical oracle we use the fact that the scaling invariance effect for the typical cell of a Cox–Voronoi tessellation that is based on a Poisson line tessellation described in Section 2.4.6, is not restricted to first moments but is also valid, for example, with respect to variances. So, if we use the number of vertices of the typical cell $\eta(\Xi_\tau^*)$ as a characteristic, we are able to state the following metamorphic relationship

$$\mathrm{Var}\,\eta(\Xi_\tau^{*(1)}) = ... = \mathrm{Var}\,\eta(\Xi_\tau^{*(k)}), \qquad (5.12)$$

where the test cases are $t_i : (\gamma, \lambda_\ell) = (\gamma^{(i)}, \lambda_\ell^{(i)})$ for $i = 1, ..., k$. The statistical MR given in (5.12) may lead to Levene–type tests (cmp. [48]) that can be used by the comparator since in this case we have a strong similarity between the test based on the

Table 5.5: Test for equality of variances by means of p–values

κ	$\tilde{\eta}(\tilde{\Xi}_\tau^*)$	$\tilde{\nu}_1(\partial\tilde{\Xi}_\tau^*)$	$\tilde{\nu}_2(\tilde{\Xi}_\tau^*)$
10	0.457	0.883	0.907
50	0.034	0.623	0.296
120	0.449	0.608	0.603

statistical MR and a test for homoscedasticity of random variables. Note that similar statistical metamorpic relations can be given if the characteristic $\eta(\Xi_\tau^*)$ is replaced by $\nu_1(\partial\Xi_\tau^*)$ or $\nu_2(\Xi_\tau^*)$, respectively. Results for statistical metamorphic tests based on these metamorphic relations are shown in Table 5.5 for three different scaling parameters $\kappa = 10$, $\kappa = 50$, and $\kappa = 120$, for eight different values of γ per scaling parameter and for a sample size of $n = 2{,}000{,}000$ per test case. It is viewable that the p–values are quite well uniformly distributed in $[0, 1]$ which should be the case if the null hypothesis holds. Hence, it is justified to say that the IUT fulfills the statistical metamorphic relations with respect to the variances of the number of vertices, the perimeter and the area of the typical cell.

5.4 Tests by Comparison to Gold–Standard

5.4.1 Basic Principle

In Sections 5.2 and 5.3 methods for testing software with random inputs and outputs have been introduced based on statistical oracles and on metamorphic relations. In this section an extension to this approach is shown, where the statistical oracle and the metamorphic relations are combined with a second implementation that is used to test the IUT. Here, the basic principle is to test two implementations against each other. One of the two implementations might be an already working implementation of the algorithm (often called the gold–standard). Due to the fact that in our regarded cases already implemented algorithms are usually not available we turn our attention to a slight modification. Implementations of algorithms for the simulation of the zero cell for the specific tessellation are used in order to test the results for the IUT that simulates the typical cell of the tessellation. Note that algorithms for the simulation of the zero cell are usually much more straight forwardly to derive and implement. Therefore, they are ideal candidates in order to be utilized as a gold–standard for testing an IUT that is simulating the typical cell of the same tessellation type. The test itself is based on statistical metamorphic relations between the typical cell and the zero cell of a tessellation and the test decision is given by a statistical oracle where the decision of the comparator is based on a statistical test. Note that from a very strict viewpoint we do not apply metamorphic testing here, since two different implementations are tested versus each other. But, since the whole procedure is very similar for a relation between two implementations compared to the testing based on a MR for a single implementation it is justified to speak about metamorphic relations also in the context of two involved implementations.

More precisely, we will look at examples where we construct two related test cases t_1 and t_2 with outputs $o(t_1)$ and $o(t_2)$. The first test case is performed with the IUT, while t_2 is performed with respect to a second implementation which is connected to the IUT via a statistical MR of their specifications for the input parameters of t_1 and t_2. This means that, assuming the IUT and the second implementation are working correctly, the outputs $o(t_1)$ and $o(t_2)$ of the IUT and the second implementation are fulfilling a statistical MR. As in Sections 5.2 and 5.3 the comparator makes the decision whether the test passes or not based on a statistical test and on a value of a test statistic that is provided by the analyser.

5.4.2 Examples

With respect to the examples considered in this section a comparison is performed between the results for the IUT that simulates the typical cell of a random tessellation and a second implementation that simulates the zero cell of the same random tessellation. The statistical MR that is applied is given by Lemma 2.9, which tells us that, given a stationary tessellation τ with typical cell Ξ_τ^* and zero cell Ξ_τ^o and a translation–invariant, non–negative and measurable function $f : \mathcal{C} \to [0, \infty)$ we have that

$$\mathbb{E}\left[f(\Xi_\tau^o)\right] = \lambda_\tau \mathbb{E}\left[f(\Xi_\tau^*)\nu_2(\Xi_\tau^*)\right].$$

In particular, if, for example, we take f to be the number of vertices η we obtain the following relationship

$$\mathbb{E}\left[\eta(\Xi_\tau^o)\right] = \lambda_\tau \mathbb{E}\left[\eta(\Xi_\tau^*)\nu_2(\Xi_\tau^*)\right], \tag{5.13}$$

which provides us with a usable statistical MR.

In the specific example the implementation for the simulation of the typical cell for a Voronoi tessellation that is induced by a Cox point process is considered (see Sections 2.4.6 and 3.1.3). For an intensity $\lambda_X = \gamma\lambda_\ell$, a given scaling invariance parameter $\kappa = \gamma/\lambda_\ell$, an intensity of the Poisson line process of $\gamma = 0.125$ and a fixed sample size of $n = 2{,}000{,}000$ the mean number of vertices for the typical cell Ξ_τ^* provided by the IUT are compared to the same characteristic provided by an implementation for the simulation of the zero cell Ξ_τ^0 of the same tessellation type. This means that if we consider a test case t_1 for our IUT, the statistical MR (5.13) induces a test case t_2 for the implementation of the simulation of the zero cell Ξ_τ^0 of the same tessellation type with the same parameter values as t_1.

Applying (5.13) we can derive a test statistic of the form

$$T = \frac{\sum_{i=1}^{n} \left(\lambda_X \tilde{\nu}_2(\tilde{\Xi}_{\tau,i}^*) \tilde{\eta}(\tilde{\Xi}_{\tau,i}^*) - \tilde{\eta}(\tilde{\Xi}_{\tau,i}^0) \right)}{\sqrt{(n-1)(S_{(*)}^2 + S_{(0)}^2)}}, \tag{5.14}$$

where

$$S_{(*)}^2 = \sum_{i=1}^{n} \left(\lambda_X \tilde{\nu}_2(\tilde{\Xi}_{\tau,i}^*) \tilde{\eta}(\tilde{\Xi}_{\tau,i}^*) - \frac{1}{n} \sum_{j=1}^{n} \lambda_X \tilde{\nu}_2(\tilde{\Xi}_{\tau,i}^*) \tilde{\eta}(\tilde{\Xi}_{\tau,j}^*) \right)^2$$

and

$$S_{(0)}^2 = \sum_{i=1}^{n} \left(\tilde{\eta}(\tilde{\Xi}_{\tau,i}^0) - \frac{1}{n} \sum_{j=1}^{n} \tilde{\eta}(\tilde{\Xi}_{\tau,j}^0) \right)^2.$$

For a large sample size n the test statistic T can be assumed to be approximately standard normal distributed which allows the construction of an asymptotic Gaussian test. Note that test statistics with respect to the mean perimeter and the mean area as well as equivalent tests can be derived analogously. Now, if the analyser computes the test statistic T based on the outputs $o(t_1)$ and $o(t_2)$ of t_1 and t_2, respectively, the comparator is able to decide whether the test passes or not based on the value of T and on the asymptotic Gaussian test. Therefore a software test for two implementations is realised based on the combination of a statistical oracle and a statistical MR.

In Table 5.6 some results of tests for a parameter value of $\kappa = 10$ and $\kappa = 120$ are displayed. The p–values are quite uniformly distributed in $[0, 1]$ which should be the case if the null hypothesis holds. Hence, if the implementation of the simulation for the zero cell of the tessellation is correct we can assume that the IUT produces correct results with respect to the statistical MR given in (5.14) and the characteristics $\mathbb{E}\eta(\Xi_\tau^*)$, $\mathbb{E}\nu_1(\partial\Xi_\tau^*)$ and $\mathbb{E}\nu_2(\Xi_\tau^*)$.

5.5 Summary of Testing Methods

In this chapter we regarded different methods for tests of software with random output, in particular for implementations of the algorithms described in Chapters 3 and 4. All of these testing methods are at least partially based on statistical oracles and therefore only provide results of testing in a statistical sense. Nevertheless, during the implementation and testing work we obtained the experience that these statistical testing methods proved to be very useful in order to detect errors in the implementations. Mostly due

Table 5.6: Tests by comparison with zero cell algorithm (fixed κ): p–values

a) $\kappa = 10$ b) $\kappa = 120$

γ	η	ν_1	ν_2
0.125	0.067	0.068	0.139
0.25	0.187	0.308	0.237
0.4	0.104	0.020	0.057
0.5	0.391	0.536	0.780
0.8	0.174	0.377	0.255
1.0	0.108	0.019	0.033
1.25	0.696	0.632	0.673
1.5	0.805	0.508	0.431

γ	η	ν_1	ν_2
0.125	0.741	0.827	0.759
0.25	0.284	0.080	0.057
0.4	0.335	0.157	0.160
0.5	0.652	0.632	0.758
0.8	0.673	0.749	0.829
1.0	0.285	0.178	0.232
1.25	0.471	0.509	0.387
1.5	0.637	0.756	0.793

to the large sample sizes that can be quite easily simulated it becomes very unlikely that an error is not detected during the testing procedures. This is very plausible since for a large sample size, the probability that the results of an implementation that behaves differently (e.g. that produces a different mean value) compared to the expected outcome, do not lead to a rejection becomes very small which can be justified by considering the construction principle of the tests that are mostly based on the central limit theorem.

Especially the tests introduced in Section 5.2 that are constructed by the application of a statistical oracle that is based on a theoretical property of a characteristic tend to detect implementation errors very rapidly. With respect to tests that are based on statistical metamorphic relations (Section 5.3) we have the impression that one has to be a bit more cautious since, although some errors are detected, sometimes they are not able to detect an error due to the fact that this error is influencing all test cases in the same manner and therefore can become simply undetectable by a method based on a particular statistical MR. Of course, this varies if different kinds of statistical metamporphic relations are considered. For example, the tests that are given by a comparison to a gold–standard implementation as described in Section 5.4 showed quite a good capability of detecting implementation errors although also in this case a the testing procedure is based partially on a statistical MR. Here, the implementations for the simulation of the zero cell for the given tessellation provide a very good tool for testing the implementation of the simulation for the typical cell due to their relative simplicity. For a more detailed discussion of the selection of good metamorphic relations see, for example, [63].

Chapter 6

Concepts of Morphological Image Analysis

The presentation of random geometrical structures like networks is often based on imaging procedures, such as microscopy or photography. The specific demands of the investigative projects determine the properties of the image, for example, the spatial resolution, the dimensionality, etc. Typically for many applications such data is given in the form of grey scale images or color images, e.g. microscopic images or photographical images. Preprocessing in this context means, apart from the removal of disturbances (noise), a segmentation of the image into meaningful regions that can be analysed later on. In order to achieve such a segmentation morphological image analysis provides a powerful set of tools. In this chapter some of these tools are introduced. In particular, after the definition of some basic notions and image filters in Section 6.1, in Section 6.2, the morphological skeletonization is introduced that is applied in Section 8.1.1 for the segmentation of image data from electron microscopy. In Section 6.3 the morphological watershed transformation is discussed which is a key component of the image segmentation algorithm given in Section 8.2.1 that is used in order to segment microscopic images of actin filament networks. Finally, in Section 6.4 some further morphological operations like pruning and merging are introduced that are also applied in the image segmentation examples given in Chapter 8. For more detailed informations on morphological image analysis in general see, for example, [41], [96], and [99].

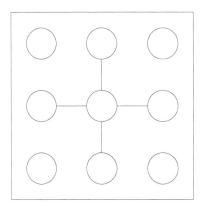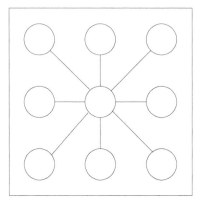

a) *4–connectivity with respect to the central pixel* b) *8–connectivity with respect to the central pixel*

Figure 6.1: Connectivity examples

6.1 Basics of Morphological Image Analysis

6.1.1 Digital Grids and Digital Images

A *digital grid* resembles a special kind of graph (cmp. Section 4.1). We usually work with a square grid $D \subset \mathbb{Z}^2$, where the vertices of D are called *pixels*. In the following we will postulate that D is finite. Then, the *size* of D is the number of points in D. We can endow the set of pixels D with a graph structure $G = (V, E)$ by setting the domain D as V and by setting a certain subset of $V \times V$ as E, thereby defining the connectivity. Typical choices with respect to E are *4–connectivity* (*4–neighbourhood*), i.e., each pixel possesses horizontal and vertical neighbours, or *8–connectivity* (*8–neighbourhood*) where each pixel is connected to its horizontal, vertical and diagonal neighbours (Figure 6.1). Another possibility is, for example, a *6–connectivity* (*6–neighbourhood*). In this case it must be differed between two different types of 6–connectivity. It can be either realised as a modification of 8–connectivity by prefering specific directions (Figure 6.2a) or alternatively one can regard grids that have a honeycomb–structure and connect neighboring combs (Figure 6.2b). By associating a non–negative cost or weight $c(e)$ to each edge $e = (p, q)$ it is possible to introduce *distances* between neighbouring pixels in a digital grid. The distance between non–neighbouring pixels of the digital grid is then defined in a canonical way as the shortest path length between them (cmp. Section 4.1). A *(digital) grey scale image* is a triple (D, E, f_g), where (D, E) is a graph (usually a

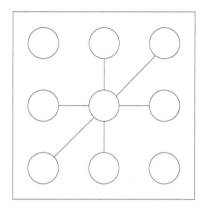 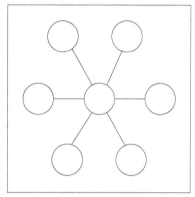

a) *6–connectivity with respect to the central*   b) *6–connectivity with respect to the central*
pixel on a square grid   *pixel on a honeycomb grid*

Figure 6.2: Different 6–connectivities

digital grid) and where $f_g : D \rightarrow \mathbb{N} \cup \{0\}$ is a mapping assigning a non–negative integer value to each $p \in D$. Often we will only refer to the mapping f_g as the grey scale image. A *binary image* f_b is an image with only two pixel values, e.g. 0 or 1. For $p \in D$ we call $f(p)$ the *grey value*. Typically the grey values in a grey scale image range between 0 and 255 but other choices are also possible. Note that the range of the grey values determines the grey scale resolution of the image which is an important factor for the amount of information contained in the image, especially with respect to contrast between different objects displayed. Another aspect is that the grey scale resolution in combination with the size of the image determines the amount of memory needed for storing the image.

A possible extension of a grey scale image in order to represent colored structures is the *RGB image* (*truecolor image*). Here, the mapping $f_c : D \rightarrow \mathbb{N}^3$ assigns to each pixel of the image three non–negative integer values that represent the intensity of the red, the green and the blue component. Hence, there exists a natural representation of a grey scale image f_g as an RGB–image f_c by putting $f_c(p) = (f_g(p), f_g(p), f_g(p))$, where $f_g(p)$ is the grey scale value for the pixel p, wheras $f_c(p)$ is the value of the RGB image at p. Note that, apart from RGB–images, various other ways of representing colored images exist, for example, HSI–images or CMY–images.

Transformations of grey scale images into binary images and of RGB images into grey scale images are not obvious and in general connected to a loss of information. A possible method for the transformation of a grey scale image into a binary image is given by the operation of thresholding that is explained in Section 6.1.2.

6.1.2   Image Filtering

A way of improving results of image segmentation is to preprocess the given image by filtering before applying morphological operations to the image. In general we can define a *filter* or a *neighbourhood operation* as an operation that, given an image, combines the pixel values of a small neighbourhood in an appropriate manner in order to yield a result that forms a new image. This new image might have a different content than the original image, but usually the aim is to preserve or extract certain characteristics from the original image. In the following some basic filters are introduced that are applied in Chapter 8. For more details on image filtering see, for example, [41].

Thresholding
One of the simplest filters is given by thresholding which serves for the purpose of binarizing a grey scale image f_g. An operator $T_{[t_l, t_u]} : \{0, ..., t_{\max}\} \rightarrow \{0, 1\}$ is a *threshold operator* if

$$[T_{[t_l, t_u]}](f_g(p)) = \mathbb{1}_{\{t_l(p) < f_g(p) < t_u(p)\}} \tag{6.1}$$

The operation of thresholding can be further divided into constant and dynamic thresholding depending on how the bounds $t_l(p)$ and $t_u(p)$ are chosen. In particular, we have a *constant thresholding* if the bounds $t_l(p)$ and $t_u(p)$ are independent of p for all $p \in D$. A *dynamic thresholding* means that $t_l(p)$ and $t_u(p)$ indeed depend on the argument p. Examples for a constant thresholding are displayed in Figure 6.3. Obviously, it is crucial for the resulting binary image how the thresholds t_l and t_u are chosen. A *plateau* or a *connected component* of grey value h is a set of pixels of constant grey value h, where for each pair of pixels (p_i, p_j) in the same plateau of grey value h it holds that there exists a path between p_i and p_j and vice versa such that each pixel along the path also has a grey value of h. We call a set T_h a *threshold set* of f_g at level h if

$$T_h = \{p \in D | f_g(p) \leq h\}. \tag{6.2}$$

Low–pass filters
In later applications the disturbances are typically of a high frequency, meaning that compared to the rather smooth grey scale values for the objects themselves, disturbances are strongly concentrated in a small region with comparable large differences in the grey scale values. Therefore, it is useful to apply low–pass filters that are able to supress such high-frequential noise and that on the other hand retain low-frequential pixel values quite well. A prominent example for such a low–pass filter is the *mean* or *average filter*

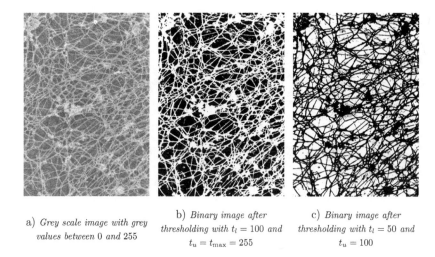

a) *Grey scale image with grey values between 0 and 255*

b) *Binary image after thresholding with $t_l = 100$ and $t_u = t_{\max} = 255$*

c) *Binary image after thresholding with $t_l = 50$ and $t_u = 100$*

Figure 6.3: Examples for thresholding

where the pixel value of the center pixel is replaced by an average with respect to a given neighbourhood. If f_g denotes the original image and f'_g is the image after applying an average filter we have that for a pixel p with its neighborhood $H(p)$

$$f'_g(p) = \frac{\sum_{i \in H(p)} w_i f_g(p_i)}{\sum_{i \in H(p)} w_i},$$

where w_i denotes the weight for the pixel p_i. Usually the weights are chosen such that they sum up to one, i.e., $\sum_{i \in H(p)} w_i = 1$ for all $p \in D$. Thereby it is achieved that the mean grey scale value of the image is not changed, where mean grey scale value means the arithemic mean of the grey values with respect to all the pixels in the image. The application of the mean filter is usually performed by using a convolution mask such as the following 3×3 matrix for an 8–neighbourhood which describes the weights for the different pixels with respect to the center pixel

$$\begin{pmatrix} 1/9 & 1/9 & 1/9 \\ 1/9 & 1/9 & 1/9 \\ 1/9 & 1/9 & 1/9 \end{pmatrix}.$$

Mean filter in general tend to blur images as you can see in Figure 6.4, where the blurring effect becomes stronger the larger the convolution mask is chosen.

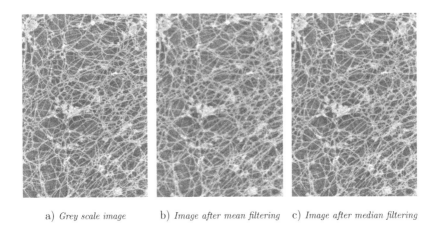

a) *Grey scale image*   b) *Image after mean filtering*   c) *Image after median filtering*

Figure 6.4: Examples for mean and median filtering

A second type of low–pass filter is given by the *median filter*, whose construction principle is similar to that of a mean filter. Instead of the average value with respect to the pixel values in a given neighbourhood, the median of the pixel values in a neighbourhood is taken as a replacement for the original value. More precisely, if f_g and f_g' denotes the grey scale image before and after application of the median filter and $H(p)$ is a given neighborhood with respect to the pixel p we obtain that

$$f_g'(p) = med_{p_i \in H(p)} f_g(p_i),$$

where $med_{p_i \in H(p)}$ denotes the median with respect to all pixels in the neigborhood of p. Such median filters are often used in order to reduce the noise in images while preserving sharp edges in the image (Figure 6.4).

A third example for a low–pass filter which is a special type of averaging filter is the Gaussian filter that is based on the Gaussian distribution for a two–dimensional random variable. Recall that we call a random variable $X = (X_1, X_2)$ with $X : \Omega \to \mathbb{R}^2$ *(bivariate) Gaussian* or *normally distributed* with mean vector $(0,0)^\top$ if its (two–dimensional) density is given by

$$f_X(x_1, x_2) = \frac{1}{2\pi \sigma_{X_1} \sigma_{X_2} \sqrt{1 - \rho^2}} \exp\left(-\frac{1}{2(1 - \rho^2)}\left(\frac{x_1^2}{\sigma_{X_1}^2} + \frac{x_2^2}{\sigma_{X_2}^2} - \frac{2\rho x_1 x_2}{(\sigma_{X_1} \sigma_{X_2})}\right)\right),$$

where ρ denotes the correlation coefficient of X_1 and X_2, and $\sigma_{X_1}^2$ and $\sigma_{X_2}^2$ are the variances of X_1 and X_2, respectively. If $\rho = 0$ and $\sigma_{X_1} = \sigma_{X_2} = \sigma_X$ we get

$$f_X(x_1, x_2) = \frac{1}{2\pi\sigma_X^2} \exp\left(-\frac{x_1^2 + x_2^2}{2\sigma_X^2}\right). \tag{6.3}$$

The basic principle of a Gaussian filter is to construct a convolution matrix that is an approximation of the bivariate Gaussian distribution given by (6.3). Theoretically we have that the density of the Gaussian distribution is non–zero everywhere, which would require an infinitely large convolution matrix (and therefore an infinitely large neighborhood) but in practice it can be assumed that the density is effectively zero more than about three standard deviations from the mean and so we are able to truncate the convolution matrix at this point. Figure 6.5 displays a sample kernel for a size of 5 and a standard deviation $\sigma_X = 1$.

The effect of Gaussian filtering is, like for other low–pass filters, to blur an image. The degree of smoothing is determined by the given standard deviation of the underlying Gaussian distribution. For larger standard deviations, of course larger convolution matrices should be used in order to be accurately represented. The result of a Gaussian filtering can be considered as a weighted average of each pixel's neighbourhood, where, compared to a filter with equal weights, the average is weighted more towards the values of the central pixels. Due to this fact, and with respect to a similarly sized average filter with equal weights, we have that the Gaussian filter in general provides more gentle smoothing in the sense that edges are better preserved. An example for an image that is processed by a Gaussian filter is shown in Figure 6.6.

Note that with respect to low-pass filters various various other techniques exist, for example, anisotropic diffusion filters ([30], [81]). In this case the filter itself is defined as a diffusion process that encourages intraregion smoothing while inhibiting interregion smoothing.

6.2   Skeletonization by Morphological Operators

The aim of morphological skeletonization algorithms is to replace given structures or objects by thin skeletons, idealistically of width one pixel, while preserving the main shape characteristics and the number of objects and holes in the structure. In particular, we use skeletonization by morphological operators in binary images. Here, eight structural elements $M_1, .., M_8$ are utilized in order to transform a given binary image into a skeleton structure. These structural elements are given by the matrices (cmp. [35])

$$M_1 = \begin{pmatrix} 0 & * & * \\ 0 & 1 & 1 \\ 0 & * & 1 \end{pmatrix}, M_2 = \begin{pmatrix} * & 1 & * \\ 0 & 1 & 1 \\ 0 & 0 & * \end{pmatrix}, M_3 = \begin{pmatrix} * & 1 & 1 \\ * & 1 & * \\ 0 & 0 & 0 \end{pmatrix}, M_4 = \begin{pmatrix} * & 1 & * \\ 1 & 1 & 0 \\ * & 0 & 0 \end{pmatrix},$$

$$\frac{1}{273}$$

1	4	7	4	1
4	16	26	16	4
7	26	41	26	7
4	16	26	16	4
1	4	7	4	1

Figure 6.5: Gaussian filter matrix for $\sigma_X = 1.0$

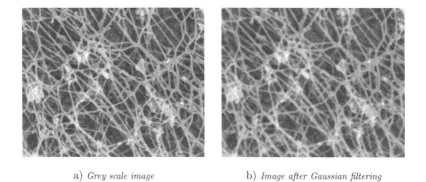

a) *Grey scale image*   b) *Image after Gaussian filtering*

Figure 6.6: Example for Gaussian filtering

$$M_5 = \begin{pmatrix} 1 & * & 0 \\ 1 & 1 & 0 \\ * & * & 0 \end{pmatrix}, M_6 = \begin{pmatrix} * & 0 & 0 \\ 1 & 1 & 0 \\ * & 1 & * \end{pmatrix}, M_7 = \begin{pmatrix} 0 & 0 & 0 \\ * & 1 & * \\ 1 & 1 & * \end{pmatrix}, M_8 = \begin{pmatrix} 0 & 0 & * \\ 0 & 1 & 1 \\ * & 1 & * \end{pmatrix},$$

where '$*$' denotes an arbitrary entry (0 or 1 for binary images). A pixel with entry 1 is set to 0 in the kth step if its corresponding matrix of neighbour entries fulfills all the conditions given by the matrix $M_{(k \bmod 8)+1}$. More precisely in the kth step the matrix $M_{(k \bmod 8)+1}$ is applied to the binary image in the following way. Let $p_{k-1}(i,j)$ denote the value of the pixel at position $(i,j) \in \mathbb{Z}^2$ after $k-1$ steps of the algorithm. Then $p_k(i,j) = 0$ if either $p_{k-1}(i,j) = 0$ or if $p_{k-1}(i,j)$ fulfills the conditions imposed by $M_{(k \bmod 8)+1}$. Otherwise $p_k(i,j) = 1$. For example, in the 12th step the structural element is given by $M_{(12 \bmod 8)+1} = M_5$. A pixel $p_{12}(i,j)$ is set to 0 if either $p_{11}(i,j) = 0$, or if $p_{11}(i,j) = 1$ and the following holds. The pixels $p_{11}(i-1,j)$ and $p_{11}(i-1,j+1)$ have to be 1 and the pixels $p_{11}(i+1,j-1)$, $p_{11}(i+1,j)$ and $p_{11}(i+1,j+1)$ must equal 0. The other three pixels in the direct neighbourhood, namely $p_{11}(i-1,j-1)$, $p_{11}(i,j-1)$ and $p_{11}(i,j+1)$ can be of arbitrary value (0 or 1). Notice that if after the k^*th step $p_{k^*}(i,j) = 0$ then $p_k(i,j) = 0$ for all $k \geq k^*$. The application of the structuring elements in a rotating fashion to the binary image is performed until $\sum_{i,j} p_k(i,j) = \sum_{i,j} p_{k+8}(i,j)$, i.e., the number of pixels that have values equal to 1 has not changed for a whole cycle of the eight structuring elements $M_{(k \bmod 8)+1}, M_{((k+1) \bmod 8)+1}, ..., M_{((k+7) \bmod 8)+1}$. Due to the assumption that the number of pixels in the binary image is finite, the algorithm terminates after a finite number of steps and due to the algorithm structure the resulting binary structures have a thickness of one pixel. An example for an application of this skeletonization algorithm to a binary image is displayed in Figure 6.7. As it is already viewable from this sample, skeletonization algorithms provide a powerful tool for the segmentation of binary images. Beyond that, after a suitable thresholding, they are also capable of segmenting certain grey scale images.

6.3 Morphological Watershed Transformation

A skeletonization by morphological operators as explained in Section 6.2 requires a binarization of the grey scale image which can lead to the loss of information. Hence, in some cases it is worthwhile to regard an alternative approach for the segmentation of grey scale images compared to thresholding and a subsequent skeletonization which is given by morphological watershed transformation. The idea of the watershed transformation originally comes from geoscience where a watershed denotes the boundary of the influence zones for two neighbouring rivers which is explained in Figure 6.8. This meaning is translated into the context of image segmentation by regarding a grey scale image f as a representation of height values, where the heights are given by the entries for each pixel of f. In the following we focus on watershed transformation based

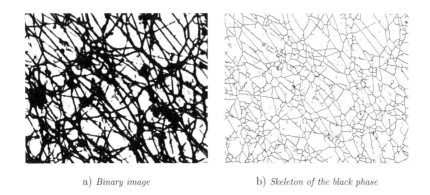

a) *Binary image*   b) *Skeleton of the black phase*

Figure 6.7: Skeletonization

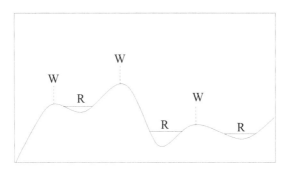

Figure 6.8: Illustration of watersheds, where R denotes the rivers and W the watersheds between them

on immersion, also called successive flooding ([110]). With regard to other watershed algorithms like marker controlled watershed see, for example, [87] and [99].

A characteristic that is important for watershed algorithms is the geodesic distance. Let $D \subset \mathbb{Z}^2$ be a (finite) digital grid and let p_1 and p_2 be two elements in D. The *geodesic distance* $d_D(p_1, p_2)$ between p_1 and p_2 within D is given by the minimum path length among all paths within D from p_1 to p_2. If D' is a subset of D we define

$$d_D(p, D') = \min_{p' \in D'} d_D(p, p')$$

for any $p \in D$. Now consider a subset $D' \subset D$ that is partitioned into k connected components D'_i, $i = 1, ..., k$. The *geodesic influence zone* of the set D'_i within D is defined as

$$iz_D(D'_i) = \{p \in D | \forall j \in \{1, ..., k\} \setminus \{i\} : d_D(p, D'_i) < d_D(p, D'_j)\}. \tag{6.4}$$

The set $IZ(D')$ is then the union of the geodesic influence zones of the connected components of D', i.e.,

$$IZ_D(D') = \bigcup_{i=1}^{k} iz_D(D'_i).$$

Now we are able to provide a definition of the watershed transformation by simulated immersion. Let f_g be a grey scale image where $h_{\min}$ and $h_{\max}$ are the minimal and maximal values of f_g. Let the grey level h increase from $h_{\min}$ to $h_{\max}$ and define a recursion in which the basins of f_g generated by the local minima of f_g are successively expanded. By X_h we denote the union of the set of basins that are computed at grey level h. Then, with respect to a connected component of the threshold set T_{h+1} there are two possibilities. It might either be a new (local) minimum or it might be an extension of a basin in X_h. In the latter case the geodesic influence zones of X_h in T_{h+1} are computed, resulting in an update X_{h+1}. If $\min_h$ denotes the set of local minima at level h we can construct the recursion as follows.

$$\left\{ \begin{array}{l} X_{h_{\min}} = \{p \in D | f(p) = h_{\min}\}, \\ X_{h+1} = \min_h \cup IZ_{T_{h+1}}(X_h), \ h \in [h_{\min}, h_{\max}], \end{array} \right. \tag{6.5}$$

The *watershed* $Wshed(f_g)$ of f_g is the complement of $X_{h_{\max}}$ in D

$$Wshed(f_g) = D \setminus X_{h_{\max}}. \tag{6.6}$$

For an example of a watershed transformation based on the above recurrence see Figure 6.9, where A, B, C, and D are labels of basins and W denotes pixels that belong to the watershed. An important property of the watershed algorithm is its dependancy on the given connectivity, meaning that for different given connectivity relations different watersheds are obtained. In practice, often a binary image is constructed that

a) Original image

2	2	3	1
1	2	3	0
2	3	3	2
0	2	1	3

b) $h = 0$

2	2	3	1
1	2	3	B
2	3	3	2
A	2	1	3

c) $h = 1$

2	2	3	B
D	2	3	B
2	3	3	2
A	2	C	3

d) $h = 2$

D	D	3	B
D	D	3	B
W	3	3	W
A	W	C	3

e) $h = 3$

D	D	W	B
D	D	W	B
W	W	W	W
A	W	C	C

Figure 6.9: Watershed transform by simulated immersion on the 8–connected grid

represents the pixels of the watershed as foreground pixels and the other pixels as background pixels. Furthermore, keep in mind that the watershed algorithm can only produce closed networks in the sense that there are no open branches or in other words dead ends in the segmented network structure.

6.4 Other Morphological Operations

In this section we describe some other morphological operations that are sometimes required in order to process and enhance the results of, e.g., a prior skeletonization or watershed transformation.

6.4.1 Processing of Line Structures

Given a binary image f and a connectivity (e.g. 8–neighbourhood) it is often necessary to classify the pixels of f into *endpoints* (having only one neighbour), *linepoints* (having exactly two neighbours) and *crosspoints* (having more than two neighbours). The connection (based on the neighbouring relationship) between two crosspoints, a crosspoint and an endpoint or two endpoints is called a *connection path* if otherwise only line points are involved. Note that often a classification using merely the number of neighbour pixels is not satisfactorily since this would result in very many pixels being classified as crosspoints and hence very short connection paths between them. Therefore criterions should be used in order to improve results (Figure 6.10). Such an improved classification should preserve the number of connected components while the number of connection paths should become minimal. A criterion to decide between different possible solutions is to regard the difference of angles for the connections of a pixel to its neighbours, where angles close to π are preferred.

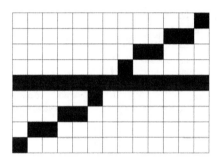

a) *Binary image*

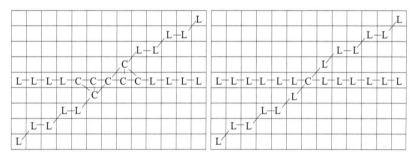

b) *Classification by number of neighbours based on 8-neighbourhood*

c) *Classification according to minimal connections while preserving number of connected components (here 1)*

Figure 6.10: Classification and its improvement, where C denotes crosspoints and L denotes linepoints

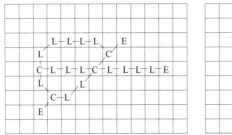 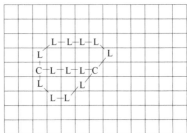

a) *Structure before pruning*   b) *Structure after iterative pruning and*
                                                    reclassification

Figure 6.11: Example of iterative pruning

6.4.2 Pruning

The operation that removes endpoints from a given classified pixel structure is called
pruning. After a subsequent reclassification of the pixels it is possible to apply this
pruning operation in an iterative fashion. If this iterative pruning is performed a sufficient number of times all dead ends of the structure are removed. An iterative pruning
procedure is often useful, e.g. if the segmented structure after applying an algorithm
like skeletonization contains some open branches although it is known that the real
structure does not contain such dead ends. For an example of an iterative pruning
operation with a subsequent reclassification of the pixels see Figure 6.11.

6.4.3 Merging

Due to the width of the natural objects that are visualised in the images it might hap-
pen that algorithms like skeletonization might produce a segmentation that gives a false
impression, especially with respect to crossings under a small angle (cmp. Figure 6.12).
Therefore, it is often useful, after a transformation of a pixel structure into a graph, to
apply an operation called merging to the resulting graph.

Transformation of a pixel structure into a graph
By assigning to connections between crosspoints a weight that is proportional to the
distance between the crosspoints, a pixel structure can be transformed into a weighted
graph or, seen from a different viewpoint, into a line segment structure, where each
pair of crosspoints that has a connection path is represented as a line segment between

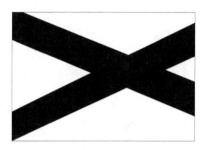 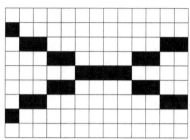

a) *Original image, two crossing lines with a certain width* b) *Result of skeletonization, a segment in the middle appears that does not reflect a structure of the original image*

Figure 6.12: Errors caused by skeletonization

them (Figures 6.13a and 6.13b).

Merging of nearby crosspoints
Given the graph structure that is the result of the transformation of the pixel structure it is possible to apply merging, where *merging* in this context means that two or more crosspoints are merged into a single one using the center of gravity of the involved crosspoints as a new crosspoint and modifying the graph structure accordingly (Figure 6.13). Typically, merging is performed for crosspoints with a distance of less than a given constant $d_{\max}$ and in an ascending sequence, meaning that crosspoints of a smaller distance are merged first.

6.5 Summary

In this chapter we introduced methods of morphological image segmentation like skeletonization and the watershed algorithm. Basically the usefulness of these methods for the applications discussed in Chapter 8 are two-fold. First of all it is a key necessity to transform the pixel-based images that represent the data into segmented structures that can be used, for example, in order to perform a statistical analysis and a later model fitting. For this purpose, the algorithms explained in Sections 6.2 and 6.3 have proven to be useful tools. Apart from that, techniques like the ones given in Sections 6.1.2 and 6.4 are able to help to improve the results by, for example, removing artifacts that are caused by disturbances in data measurement (noise, preparation artifacts, etc.).

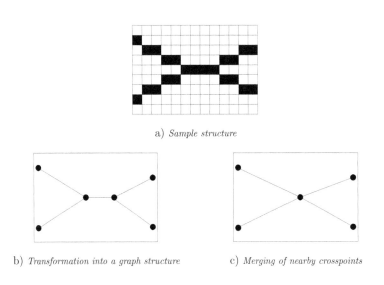

a) *Sample structure*

b) *Transformation into a graph structure*   c) *Merging of nearby crosspoints*

Figure 6.13: Transformation and improvement of the structure

Chapter 7

Fitting Random Tessellation Models to Network Structures

In this chapter an algorithm is introduced that can be used in order to fit random tessellation models to network structures, such as, for example, segmented microscopic images of the cytoskeleton or urban infrastructures. The algorithm is based on the construction of a distance function that is able to determine the distance between a theoretical tessellation model and a given data set that represents a network structure. Thereby, given a set of possible tessellation models, an optimal model as well as corresponding optimal model parameters can be determined. Data that is representing network structures in such a context means that the data is more or less represented by line segments that are connected to each other at their respective endpoints and that form cells or meshes. Often, in order to obtain such structures from observed data it is necessary to apply preprocessing techniques like the ones explained in Chapter 6. The fields of applications where such a modelling approach of representing network structures by random tessellations is useful are numerous, for example, in biology, material science, medicine and telecommunication. In Chapter 8 some applications of the model fitting algorithm introduced here with respect to cytoskeletal structures of cancer cells are discussed. Additional information about the model fitting algorithm, including examples for simulated input data and an application to network structures coming from the field of telecommunication, can be found in [32]. Mathematical definitions of the tessellation models used are provided in Section 2.4.

Table 7.1: Considered characteristics with respect to unit area

Characteristic	Meaning
λ_1	Number of vertices (nodes)
λ_2	Number of edges
λ_3	Number of cells (meshes)
λ_4	Total length of edges

7.1   Characteristics of Input Data and Estimation of Global Characteristics

7.1.1   Input Data Characteristics

In the following we assume that (the possibly preprocessed) input data is given in a rectangular sampling window W. A first step for a later model fitting algorithm is to extract certain characteristics that describe the spatial–geometric features of the input data that represents a network structure (cmp. Figure 7.1). In particular, we consider the characteristics $\lambda_1, ..., \lambda_4$, which represent the mean number of vertices, the mean number of edges, the mean number of cells or meshes and the mean total length of edges, respectively, always with respect to the unit area (cmp. Table 7.1) Note that these characteristics $\lambda_1, ..., \lambda_4$ can be considered as global characteristics from the purpose of model fitting. Besides such global characteristics one could also think of local characteristics which refer to single cells of the network structure like the mean area or the mean perimeter per cell. However, it turns out that in general the local characteristics are less useful due to the facts that often unbiased estimators are not obvious and that they do not reflect the network structure as a whole ([38]). Hence, in the following it is focused on the vector of global characteristics

$$\lambda = (\lambda_1, ..., \lambda_4)^\top. \tag{7.1}$$

7.1.2   Unbiased Estimation

In order to estimate the vector λ given in (7.1) a vector of estimators

$$\widehat{\lambda} = (\widehat{\lambda}_1, ..., \widehat{\lambda}_4)^\top \tag{7.2}$$

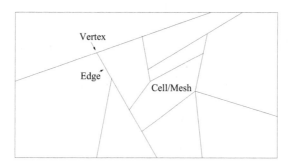

Figure 7.1: A given network structure with vertices, edges, and cells (meshes)

is needed, where each entry of the vector in (7.2) is a suitable estimator for the corresponding entry in (7.1). In the course of this thesis the vector of estimators

$$\widehat{\lambda} = \frac{1}{|W|}(n_\nu, n_e, n_c, l_e)^\top \tag{7.3}$$

is used, where n_ν is the number of vertices contained within the sampling window W. In order to get an estimate for λ_2, the mean number of edges per unit area, we use n_e, the number of edges whose lexicographically smaller end point is located in W, and divide afterwards by the area of W. An estimator for λ_3, the mean number of cells per unit area, is given by n_c, the number of cells whose lexicographically smallest vertice is located in W, divided once more by the area of W. Finally, λ_4, the mean total length of edges per unit area can be estimated by $l_e/|W|$, where l_e measures the total length of the edge–set contained in W.

7.2 Fitting Algorithm

7.2.1 Distance Functions

In order to enable a later comparison of the estimated vector of characteristics to a corresponding vector of calculated mean values for a given theoretical tessellation model a distance function has to be considered. In this thesis we regard the relative Euclidean distance function given by

$$d_{re}(x,y) = \left(\sum_{i=1}^{m} \left(\frac{x_i - y_i}{x_i} \right)^2 \right)^{1/2}, \tag{7.4}$$

where $x = (x_1, \ldots, x_m)^\top \in \mathbb{R}^m$ and $y = (y_1, \ldots, y_m)^\top \in \mathbb{R}^m$ denote two vectors with m (in our case 4) real–valued entries. Apart from the relative Euclidean distance function other choices are possible like the absolute Euclidean distance function given by

$$d_e(x, y) = \left(\sum_{i=1}^{m} (x_i - y_i)^2 \right)^{1/2} ,$$ (7.5)

the relative absolute value distance function $d_{ra}(x, y)$ and the absolute absolute value distance function $d_a(x, y)$ given by

$$d_{ra}(x, y) = \sum_{i=1}^{m} \left| \frac{x_i - y_i}{x_i} \right| ,$$ (7.6)

and by

$$d_a(x, y) = \sum_{i=1}^{m} |x_i - y_i| ,$$ (7.7)

respectively. Finally, one could, for example, also consider the relative maximum norm distance function $d_{rm}(x, y)$ and the absolute maximum norm distance function $d_m(x, y)$ that are given by

$$d_{rm}(x, y) = \max_{i=1,\ldots,m} \left| \frac{x_i - y_i}{x_i} \right| ,$$ (7.8)

and by

$$d_m(x, y) = \max_{i=1,\ldots,m} |x_i - y_i| ,$$ (7.9)

respectively.

Note that in general relative distance functions are preferable compared to absolute ones. They are scaling invariant, i.e., they are in a sense independent from the scale the data is measured on, and each component of the vector has the same influence on the result of the fitting procedure, which might not be true for absolute distance functions. Thus, absolute distance functions can be strongly influenced by a single component with a possibly extreme value. Obviously such an effect is not very convenient. Other choices (than the relative Euclidean) of relative distance functions are of course possible, but in general our experience is that the results for different relative distance functions did not vary much with respect to the choice of the optimal model as well as the values of the corresponding model parameters. Hence, we will concentrate on the relative Euclidean distance in the following. An important property of the relative distance functions that has to be noted is that they are not symmetric in their arguments x and y anymore. Therefore, some attention has to be spend in order to always use the same reference argument.

7.2.2   Optimal Model Choice

The fitting algorithm leading to an optimal tessellation model with optimal parameters can be summarized as follows. Given a pool of possible tessellation models considered and an input image, the vector of characteristics $\lambda = (\lambda_1, ..., \lambda_4)^\top$ is estimated leading to estimates $\widehat{\lambda} = (\widehat{\lambda}_1, ..., \widehat{\lambda}_4)^\top$ as described in Section 7.1. Using the relative Euclidean distance function introduced in (7.4) for each of the possible tessellation models separately a relative distance function

$$f_{model}(\theta) = \left(\sum_{i=1}^{4} \left(\frac{\widehat{\lambda}_i - \lambda_i^{model}(\theta)}{\widehat{\lambda}_i} \right)^2 \right)^{1/2} \tag{7.10}$$

can be constructed, where θ is the vector of corresponding model parameters (for example, in the case of a one–fold iterated tessellation model without Bernoulli thinning that can be described by two parameters γ_0 and γ_1, we have that $\theta = (\gamma_0, \gamma_1)$). The values $\lambda_i^{model}(\theta)$ are the theoretical model characteristics depending on the choice of the model and of the model parameter vector θ. Now, for each model separately an optimal parameter vector θ_{model}^* is determined that minimizes $f_{model}(\theta)$ for the given model. Finally, a model is considered optimal among all possible models if the optimal value $f_{model}(\theta_{model}^*)$ is minimal with respect to all models. In other words we have a two step mechanism. First, for each model seperately an optimal parameter vector is determined and afterwards the optimal model is determined by comparing the optimal values for each model. For approaches with respect to the solution of the minimization problem based on the functional given in (7.10) see [92] and [106]. Note that the fitting procedure can be slightly modified if instead of a single input image a set of n images with $n > 1$ is given, where the images are independent of each other and sampled from an identically distributed population. In this case, instead of considering the estimates $\widehat{\lambda} = (\widehat{\lambda}_1, ..., \widehat{\lambda}_4)^\top$, a vector of mean characteristics $\overline{\widehat{\lambda}} = ((\overline{\widehat{\lambda}}_1, \overline{\widehat{\lambda}}_2, \overline{\widehat{\lambda}}_3, \overline{\widehat{\lambda}}_4)^\top$ is estimated from the n samples by

$$\overline{\widehat{\lambda}}_i = \frac{1}{n} \sum_{j=1}^{n} \widehat{\lambda}_{ij}, \tag{7.11}$$

where $i = 1, ..., 4$ and where $\widehat{\lambda}_{ij}$ is the estimated ith characteristic for the jth sample. With respect to the relative distance function $f_{model}(\theta)$ the estimated vector $\widehat{\lambda}$ is then replaced by the estimated mean vector $\overline{\widehat{\lambda}}$.

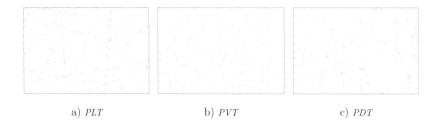

a) *PLT*   b) *PVT*   c) *PDT*

Figure 7.2: Sample realisations for basic tessellations of Poisson–type

7.3   Possible Tessellation Models Considered

In the examples that will be presented in Chapter 8 we will regard certain sets of possible tessellation models. All of them are based on three basic tessellations of Poisson–type, the (stationary) Poisson line tessellation (PLT), the (stationary) Poisson–Voronoi tessellation (PVT), and the (stationary) Poisson–Delaunay tessellation (PDT) (Figure 7.2). In particular, apart from the three basic tessellation models mentioned, we will consider one–fold nestings (including a Bernoulli thinning) of these three basic models and one–fold superpositions of the basic models (cmp. Section 2.4 and Figures 7.3–7.5). Note that for nestings we end up with nine different combinations of the three basic models, whereas for superpositions we have only six different models since in this case there is no hierarchical order between the two layers and therefore, as already mentioned in Section 2.4.8, for example a PVT/PLT superposition has the same distribution as a PLT/PVT superposition. Furthermore, note that an n–fold superposition of PLT tessellations with parameters $\gamma_1,...,\gamma_n$ is equal in distribution to a basic PLT tessellation with parameter $\gamma = \sum_{i=1}^{n} \gamma_i$. A similar relationship is not given with respect to the two other tessellation models considered, the Poisson–Voronoi tessellation and the Poisson–Delaunay tessellation. This means that, for example, an n–fold superposition of Poisson–Voronoi tessellations can not be expressed as a basic Poisson–Voronoi tessellation with respect to its distribution.

With respect to the tessellation models regarded, namely basic tessellation models as well as one–fold nestings and one–fold superpositions, we are able to derive theoretical formulae for the components of the vector of characteristics $\lambda = (\lambda_1, ..., \lambda_4)^\top$ defined in (7.1) (Tables 7.2–7.4). Note that these theoretical formulae for specific stationary tessellation models are based on more general formulae for stationary iterated tessellations that can be found in [54], [67], and [90].

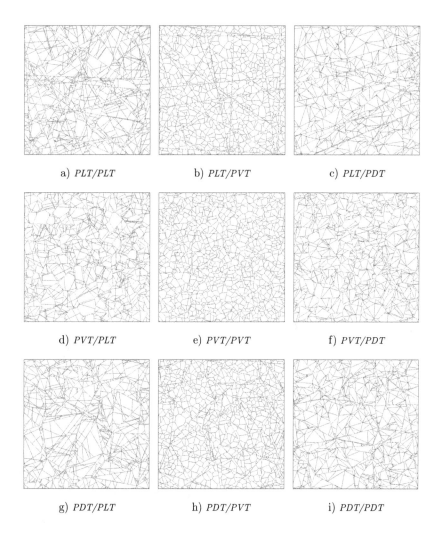

a) *PLT/PLT*   b) *PLT/PVT*   c) *PLT/PDT*

d) *PVT/PLT*   e) *PVT/PVT*   f) *PVT/PDT*

g) *PDT/PLT*   h) *PDT/PVT*   i) *PDT/PDT*

Figure 7.3: Realisations of one–fold nestings with PLT, PVT and PDT as basic tessellations shown in red

Table 7.2: Values of $\lambda_1, \ldots, \lambda_4$ for a given basic tessellation with parameter γ

Tessellation	λ_1	λ_2	λ_3	λ_4
PLT	$\frac{1}{\pi}\gamma^2$	$\frac{2}{\pi}\gamma^2$	$\frac{1}{\pi}\gamma^2$	γ
PVT	2γ	3γ	γ	$2\sqrt{\gamma}$
PDT	γ	3γ	2γ	$\frac{32}{3\pi}\sqrt{\gamma}$

Table 7.3: Mean value formulae for one–fold nestings with Bernoulli thinning

	PLT/PLT	PLT/PVT	PLT/PDT
λ_1	$\frac{1}{\pi}\gamma_0^2 + \frac{1}{\pi}p_B\gamma_1^2 + \frac{4}{\pi}p_B\gamma_0\gamma_1$	$\frac{1}{\pi}\gamma_0^2 + 2p_B\gamma_1 + \frac{8}{\pi}p_B\gamma_0\sqrt{\gamma_1}$	$\frac{1}{\pi}\gamma_0^2 + p_B\gamma_1 + \frac{128}{3\pi^2}p_B\gamma_0\sqrt{\gamma_1}$
λ_2	$\frac{2}{\pi}\gamma_0^2 + \frac{2}{\pi}p_B\gamma_1^2 + \frac{6}{\pi}p_B\gamma_0\gamma_1$	$\frac{2}{\pi}\gamma_0^2 + 3p_B\gamma_1 + \frac{12}{\pi}p_B\gamma_0\sqrt{\gamma_1}$	$\frac{2}{\pi}\gamma_0^2 + 3p_B\gamma_1 + \frac{64}{\pi^2}p_B\gamma_0\sqrt{\gamma_1}$
λ_3	$\frac{1}{\pi}\gamma_0^2 + \frac{1}{\pi}p_B\gamma_1^2 + \frac{2}{\pi}p_B\gamma_0\gamma_1$	$\frac{1}{\pi}\gamma_0^2 + p_B\gamma_1 + \frac{4}{\pi}p_B\gamma_0\sqrt{\gamma_1}$	$\frac{1}{\pi}\gamma_0^2 + 2p_B\gamma_1 + \frac{64}{3\pi^2}p_B\gamma_0\sqrt{\gamma_1}$
λ_4	$\gamma_0 + p_B\gamma_1$	$\gamma_0 + 2p_B\sqrt{\gamma_1}$	$\gamma_0 + \frac{32}{3\pi}p_B\sqrt{\gamma_1}$

	PVT/PLT	PVT/PVT	PVT/PDT
λ_1	$\frac{1}{\pi}p_B\gamma_1^2 + 2\gamma_0 + \frac{8}{\pi}p_B\gamma_1\sqrt{\gamma_0}$	$2(\gamma_0 + p_B\gamma_1) + \frac{16}{\pi}p_B\sqrt{\gamma_0\gamma_1}$	$2\gamma_0 + p_B\gamma_1 + \frac{256}{3\pi^2}p_B\sqrt{\gamma_0\gamma_1}$
λ_2	$\frac{2}{\pi}p_B\gamma_1^2 + 3\gamma_0 + \frac{12}{\pi}p_B\gamma_1\sqrt{\gamma_0}$	$3(\gamma_0 + p_B\gamma_1) + \frac{24}{\pi}p_B\sqrt{\gamma_0\gamma_1}$	$3(\gamma_0 + p_B\gamma_1) + \frac{128}{\pi^2}p_B\sqrt{\gamma_0\gamma_1}$
λ_3	$\frac{1}{\pi}p_B\gamma_1^2 + \gamma_0 + \frac{4}{\pi}p_B\gamma_1\sqrt{\gamma_0}$	$\gamma_0 + p_B\gamma_1 + \frac{8}{\pi}p_B\sqrt{\gamma_0\gamma_1}$	$\gamma_0 + 2p_B\gamma_1 + \frac{128}{3\pi^2}p_B\sqrt{\gamma_0\gamma_1}$
λ_4	$p_B\gamma_1 + 2\sqrt{\gamma_0}$	$2(\sqrt{\gamma_0} + p_B\sqrt{\gamma_1})$	$2\sqrt{\gamma_0} + \frac{32}{3\pi}p_B\sqrt{\gamma_1}$

	PDT/PLT	PDT/PVT	PDT/PDT
λ_1	$\frac{1}{\pi}p_B\gamma_1^2 + \gamma_0 + \frac{128}{3\pi^2}p_B\gamma_1\sqrt{\gamma_0}$	$2p_B\gamma_1 + \gamma_0 + \frac{256}{3\pi^2}p_B\sqrt{\gamma_1\gamma_0}$	$\gamma_0 + p_B\gamma_1 + \frac{4096}{9\pi^3}p_B\sqrt{\gamma_0\gamma_1}$
λ_2	$\frac{2}{\pi}p_B\gamma_1^2 + 3\gamma_0 + \frac{64}{\pi^2}p_B\gamma_1\sqrt{\gamma_0}$	$3(p_B\gamma_1 + \gamma_0) + \frac{128}{\pi^2}p_B\sqrt{\gamma_1\gamma_0}$	$3(\gamma_0 + p_B\gamma_1) + \frac{2048}{9\pi^3}p_B\sqrt{\gamma_0\gamma_1}$
λ_3	$\frac{1}{\pi}p_B\gamma_1^2 + 2\gamma_0 + \frac{64}{3\pi^2}p_B\gamma_1\sqrt{\gamma_0}$	$p_B\gamma_1 + 2\gamma_0 + \frac{128}{3\pi^2}p_B\sqrt{\gamma_1\gamma_0}$	$2(\gamma_0 + p_B\gamma_1) + \frac{2048}{9\pi^3}p_B\sqrt{\gamma_1\gamma_0}$
λ_4	$p_B\gamma_1 + \frac{32}{3\pi}\sqrt{\gamma_0}$	$2p_B\sqrt{\gamma_1} + \frac{32}{3\pi}\sqrt{\gamma_0}$	$\frac{32}{3\pi}\left(\sqrt{\gamma_0} + p_B\sqrt{\gamma_1}\right)$

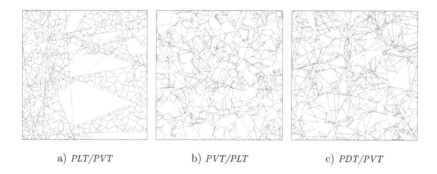

a) *PLT/PVT*   b) *PVT/PLT*   c) *PDT/PVT*

Figure 7.4: Realisations of one–fold nested tessellations with Bernoulli thinning ($p_B = 0.8$)

Table 7.4: Mean value formulae for one–fold superpositions

	PLT/PLT	PLT/PVT	PLT/PDT
λ_1	$\frac{1}{\pi}(\gamma_0 + \gamma_1)^2$	$\frac{1}{\pi}\gamma_0^2 + 2\gamma_1 + \frac{4}{\pi}\gamma_0\sqrt{\gamma_1}$	$\frac{1}{\pi}\gamma_0^2 + \gamma_1 + \frac{64}{3\pi^2}\gamma_0\sqrt{\gamma_1}$
λ_2	$\frac{2}{\pi}(\gamma_0 + \gamma_1)^2$	$\frac{2}{\pi}\gamma_0^2 + 3\gamma_1 + \frac{8}{\pi}\gamma_0\sqrt{\gamma_1}$	$\frac{2}{\pi}\gamma_0^2 + 3\gamma_1 + \frac{128}{3\pi^2}\gamma_0\sqrt{\gamma_1}$
λ_3	$\frac{1}{\pi}(\gamma_0 + \gamma_1)^2$	$\frac{1}{\pi}\gamma_0^2 + \gamma_1 + \frac{4}{\pi}\gamma_0\sqrt{\gamma_1}$	$\frac{1}{\pi}\gamma_0^2 + 2\gamma_1 + \frac{64}{3\pi^2}\gamma_0\sqrt{\gamma_1}$
λ_4	$\gamma_0 + \gamma_1$	$\gamma_0 + 2\sqrt{\gamma_1}$	$\gamma_0 + \frac{32}{3\pi}\sqrt{\gamma_1}$

	PVT/PVT	PVT/PDT	PDT/PDT
λ_1	$2\gamma_0 + 2\gamma_1 + \frac{8}{\pi}\sqrt{\gamma_0\gamma_1}$	$2\gamma_0 + \gamma_1 + \frac{128}{3\pi^2}\sqrt{\gamma_0\gamma_1}$	$\gamma_0 + \gamma_1 + \frac{2048}{9\pi^3}\sqrt{\gamma_0\gamma_1}$
λ_2	$3\gamma_0 + 3\gamma_1 + \frac{16}{\pi}\sqrt{\gamma_0\gamma_1}$	$3\gamma_0 + 3\gamma_1 + \frac{256}{3\pi^2}\sqrt{\gamma_0\gamma_1}$	$3\gamma_0 + 3\gamma_1 + \frac{4096}{9\pi^3}\sqrt{\gamma_0\gamma_1}$
λ_3	$\gamma_0 + \gamma_1 + \frac{8}{\pi}\sqrt{\gamma_0\gamma_1}$	$\gamma_0 + 2\gamma_1 + \frac{128}{3\pi^2}\sqrt{\gamma_0\gamma_1}$	$2\gamma_0 + 2\gamma_1 + \frac{2048}{9\pi^3}\sqrt{\gamma_0\gamma_1}$
λ_4	$2\sqrt{\gamma_0} + 2\sqrt{\gamma_1}$	$2\sqrt{\gamma_0} + \frac{32}{3\pi}\sqrt{\gamma_1}$	$\frac{32}{3\pi}\sqrt{\gamma_0} + \frac{32}{3\pi}\sqrt{\gamma_1}$

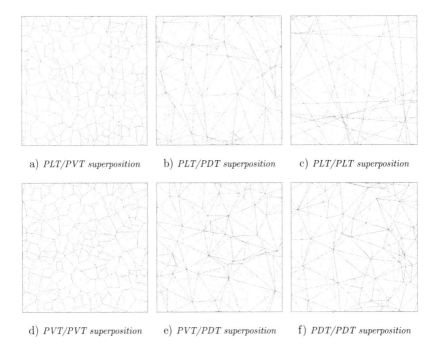

a) *PLT/PVT superposition*   b) *PLT/PDT superposition*   c) *PLT/PLT superposition*

d) *PVT/PVT superposition*   e) *PVT/PDT superposition*   f) *PDT/PDT superposition*

Figure 7.5: Realisations of one–fold superpositions with PLT, PVT and PDT as basic tessellations shown in red

Chapter 8

Applications to Electron Microscopy Images of Cytoskeletal Networks

In this chapter two applications for the methods introduced in Chapters 6 and 7 are discussed. Both of them deal with the analysis of electron microscopy images of cytoskeletal network structures in tumor cells, the first application with keratin filament networks, the second with actin network structures.

The cytoskeleton is a three–dimensional cellular scaffold within the cytoplasm. It represents a dynamic structure that is responsible for the maintenance of cell shape and for cell motility and that plays a key–role in intracellular transport as well as in cell division. The filaments or fibres of the cytoskeleton represents protein polymers. The primary types of filaments comprising the cytoskeleton are microfilaments, intermediate filaments and microtubules ([1], [49]).

The microtubules are cylindric in shape with a diameter of about 25 nm which makes them the thickest of all cytoskeletal filaments. They are formed by 13 profilaments that, in turn, are polymeres of α– and β–tubulin. Microtubules represent a scaffold that determine cell shape and provide a set of tracks for cell organelles and vesicles to move on ([76]).

The intermediate filaments have a diameter of 8 to 11 nm. Different intermediate filaments are

- keratin filaments, found, for example, in skin cells, hairs and nails,

- vimentin filaments, being the common structural support of many cells,

- neurofilaments in neural cells, and

- lamin filaments, giving structural support to the nuclear envelope of the cell.

Intermediate filaments represent the more stable part of the cytoskeleton due to the fact that they are relatively strongly bound to each other ([79]).

Microfilaments (actin) are fine and thread–like protein filaments with a diameter of about 3–6 nm, thereby making them the thinest type of filaments in the cytoskeleton. An actin filament is composed of two actin chains oriented in a helicoidal shape. Mostly they are concentrated near the plasma membrane in the so–called lamellipodium, where they keep the cell shape from cytoplasmic protuberances and where they participate in cell–to–cell or cell–to–matrix junctions as well as in the transduction of signals. They also possess a certain importance with respect to cytokinesis and muscular contraction ([82]). For more information on the cytoskeleton in general and on other aspects of cell biology see, for example, [1] and [49].

8.1 Statistical Analysis of Keratin Filament Structures

In the first example from cell biology we regard cytoskeletal structures that are formed by keratin filaments. Keratins belong to the group of intermediate filaments and are expressed in epithelial cells. The keratin cytoskeleton plays an important role for biophysical properties of the cell. In this section the remodelling of keratin filament networks in pancreatic cancer cells is investigated, where the remodelling is caused by the addition of transforming growth factor α (TGFα), which is involved in cancer cell progression. The image segmentation procedure that leads to the detection of the network structures from the given electron microscopy images is described in Section 8.1.1. In Section 8.1.2 a basic quantitative analysis of the detected segments with respect to their lengths and their orientations is performed whose results show that TGFα indeed causes a remodelling of the keratin network in the cytoskeleton. In Section 8.1.3 finally, this remodelling is investigated in more detail by fitting one–fold nested tessellation models to the given scenarios. Here, the remodelling manifests itself by having different optimal nested tessellation models with respect to the structures after TGFα incubation compared to a control group. The results of Section 8.1 are also partially documented in [8] and [9].

8.1.1 Image Acquisition and Segmentation

Human Panc–1 pancreatic cancer cells (American Type Culture Collection, Manassas, VA) are grown on glass chamber slides. The cells were treated with 100 ng/ml TGFα (R&D Systems, Minneapolis, MN) for 30 min. After washing the cells with phosphate–buffered saline (PBS), an extraction solution (1% of Triton X–100, 2.2% PEG (molecular weight, 35 kDa), 50 mM imidazole, 50 mM potassium chloride, 0.5 mM magnesium chloride, 0.1 mM EDTA and 1 mM EGTA, pH 6.9) is added for 20 min at 4 C. Cells are rinsed four times with PBS and fixed with 4.0% formaldehyde (ultrapure EM grade, methanol free, Polysciences, Eppelheim, Germany) in 0.1 M cacodylate buffer (pH 7.3) for 10 min at room temperature. After rinsing with PBS, the slides are gradually dehydrated in propanol (30%, 50%, 70%, 90% and two times in 100% for 5 min each step) and finally subjected to critical point drying using carbon dioxide as transitional medium (Critical Point Dryer CPD 030, Bal–Tec, Principality of Liechtenstein). After drying the cover slips are cutted to fit onto the Hitachi S–5200 specimen holders (5 x 8 mm) using a home made diamond cutter. The cutted glass slides are then mounted on the Hitachi holders using double sticking tape and liquid silver paint to improve electrical conductivity. Samples are subsequently rotary coated with 3 nm of platinum–carbon by electron beam evaporation using a Bal–Tec Baf 300 freeze etching device (Bal–Tec, Principality of Liechtenstein) and imaged with a Hitachi S–5200 in–lens SEM (Tokyo, Japan) at an accelerating voltage of 4 kV using the secondary electron signal. Note that this method depletes actin filaments and microtubules and preserves keratin filaments ([11]).

To analyse the filamentous structures of the keratin network for Panc1 human pancreatic cancer cells sample regions are imaged at a primary magnification of $35000x$ (pixel size 2.63 nm). In order to enable a subsequent analysis with respect to two–dimensional images and structures, sample regions are chosen from the subcortical compartment of the cells, directly adjacent to the cytoplasmic membrane. This procedure results in sample images that contain a thin (almost two–dimensional) layer of filaments. Altogether 15 sample images are analysed, 7 from cells after an incubation with TGFα and 8 from untreated cells.

The aim of the image segmentation algorithm that is applied, is to transform the given grey scale images into a graph or network structure. This network structure should contain segments that are formed by the filaments of the original keratin network. Due to measurement artifacts and noisy images such an image segmentation can of course never be perfect. Figure 8.1 displays the different image segmentation steps that are applied. As a first step the grey scale image f_g is binarized into an image f_b using a constant threshold operator $T_{[t_l, t_u]}$ (Section 6.1.2) with thresholds $t_l = \bar{f}_g$ and $t_u = f_g^{\max} + 1$, where $\bar{f}_g$ is the mean and $f_g^{\max}$ is the maximal grey scale value within f_g, respectively. The resulting binary image f_b is afterwards skeletonized by applying the skeletonization algorithm explained in Section 6.2 thereby preserving the number of connected objects and connectivity relations. After a classification of the pixels that belong to the skeleton (Section 6.4.1) it is possible to first prune iteratively the skeleton in order to remove, for example, dead ends or disturbances caused by measurement artifacts (Section 6.4.2). After the pixel classification and the iterative pruning, a structure is obtained that consists of line segments. Finally, these line segment structures are modified by merging nearby crossings to obtain line segment structures that can be statistically analysed in the following (Section 6.4.3). The maximal merging distance is chosen as $d_{\max} = 8$ pixels ($\approx 20\ nm$), thereby reflecting quite well twice the thickness of the filaments.

Figure 8.2 shows some visual results of the image segmentation algorithm explained above with respect to three different sample images. Problems are visible, where spaces between filaments are extremely narrow, in other words where filaments are touching each other, or naturally where filament connections can hardly differentiated, even by the human eye. From these examples it is also deduceable that an assumption of piecewise linearity of connections between crosspoints is a reasonable choice since most of the connections are quite small and longer connections are approximately linear.

8.1.2 Analysis of Filament Lengths and Orientations

As a first approach in order to analyse the effects of TGFα incubation on keratin cytoskeleton networks, the lengths and the orientations of the line segments resulting from the application of the image segmentation algorithm described in Section 8.1.1 are investigated.

Estimation of segment lengths and orientation angle
Segment lengths are estimated by using the Euclidean distance between the two endpoints of the segment. Segment lengths are taken into account if the typical point of the segment, in this case the lexicographically smaller endpoint, is located inside the sampling window. Note that with respect to the measurement of segment lengths the sampling window has to be chosen sufficiently small with respect to the observed image in order to avoid edge effects caused by the inability to determine whole lengths of relevant segments. With regard to the analysis of segment lengths the sampling window is chosen to be quadratic with a side length of 500 pixels which corresponds to approximately 1320 nm. It is centred at the midpoint of the image.

Orientations of segments are determined as the angle between the segment and the horizontal axis. Note that in this particular example we have axial data given. This means that a line segment has two possible directions of which neither is preferred. Therefore, the interval for possible angles can be reduced to $[-\pi/2, \pi/2]$ by simply taking the direction of the segment which has an angle that lies in this interval. Only

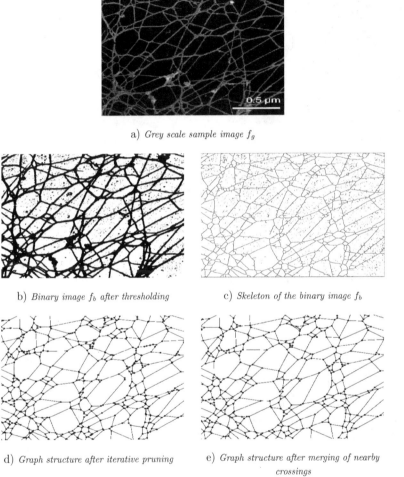

a) *Grey scale sample image f_g*

b) *Binary image f_b after thresholding*   c) *Skeleton of the binary image f_b*

d) *Graph structure after iterative pruning*   e) *Graph structure after merging of nearby crossings*

Figure 8.1: Intermediate steps of image segmentation

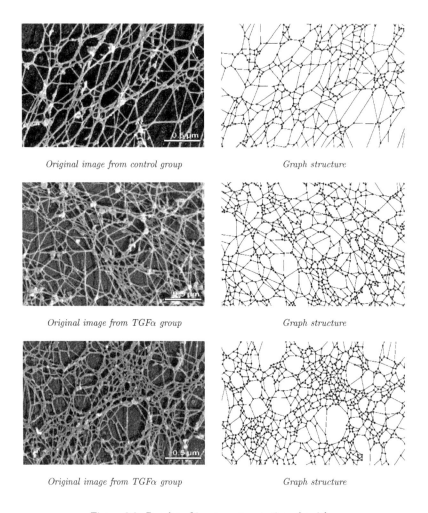

Original image from control group   *Graph structure*

Original image from TGFα group   *Graph structure*

Original image from TGFα group   *Graph structure*

Figure 8.2: Results of image segmentation algorithm

segments whose lexicographically smaller endpoint is located in the sampling window are considered. With respect to orientation analysis it is higly advisable to choose a circular sampling window in order to avoid anisotropy effects purely caused by the choice of a sampling window that favors specific directions. For the orientation analysis the sampling window is chosen to be circular with a radius of 300 pixels that corresponds to approximately 790 nm.

Orientation characteristics

Whereas characteristics for the lengths and numbers of segments are straightforward, it is more difficult to come up with characteristics that reflect the degree of anisotropy of the segments. A possible characteristic for an orientation analysis is the circular standard deviation that is closely related to the circular standard variance. In the following this characteristic will be explained. For an overview with respect to orientation analysis have a look at, for example, [6] and [66].

Let $z_1, ..., z_n$ be a sample of line segment angles measured with respect to the horizontal axis and having corresponding unit vectors $u_1, ..., u_n$ in $\mathbb{R}^2$. Then, the *mean direction* $\bar{z}$ of $z_1, ..., z_n$ is the direction of the centre of mass $\bar{u}$ of $u_1, ..., u_n$, where $\bar{u}$ can be described in Cartesian coordinates by

$$\bar{u} = (\bar{C}, \bar{S}), \tag{8.1}$$

with

$$\bar{C} = \frac{1}{n} \sum_{j=1}^{n} \cos z_j \tag{8.2}$$

and

$$\bar{S} = \frac{1}{n} \sum_{j=1}^{n} \sin z_j, \tag{8.3}$$

respectively. Therefore, given that $\bar{R} > 0$, we obtain that the mean direction $\bar{z}$ is the solution of the system of equations

$$\bar{C} = \bar{R} \cos \bar{z}, \quad \bar{S} = \bar{R} \sin \bar{z}, \tag{8.4}$$

where the *mean resultant length* $\bar{R}$ is defined by

$$\bar{R} = (\bar{C}^2 + \bar{S}^2)^{1/2}. \tag{8.5}$$

The *sample circular variance* V can now be defined as

$$V = 1 - \bar{R}. \tag{8.6}$$

It is used as a function of dispersion because if the n observed angles $z_1, ..., z_n$ are tightly clustered about the mean direction $\bar{z}$, the sample circular variance V will be almost 0. If, on the other hand, the directions are widely dispersed, we get that V is near 1. Analogously to the standard deviation in the linear case, it is possible to define a *circular standard deviation* σ as

$$\sigma = (-2\log(1-V))^{1/2}. \tag{8.7}$$

It has similar dispersion function properties as V, therefore, if $z_1 = (z_{11}, ..., z_{1n_1}), ...,$ $z_m = (z_{m1}, ..., z_{mn_m})$ denote m samples of line segment angles with sizes $n_1, ..., n_m$, respectively, we investigate the sample circular standard deviations $\sigma_1, ..., \sigma_m$ in order to obtain information with respect to orientation distributions.

Apart from the information obtained by the circular standard deviation it is often necessary to perform formal statistical tests on isotropy or in other words on uniform distribution of the angles. Such a test is, for example, given by Kuiper's test ([66], pp. 99–103). Other choices of such tests as, for example, Rayleigh's test or Watson's U^2 test ([66], pp. 94–98 and pp. 103–105) are of course also possible.

Results for the number of segments and segment lengths
Figure 8.3 shows for both regarded groups, TGFα–incubated as well as untreated control cells, a sample histogram of segment lengths for a specific image and a cumulated histogram for all segment lengths with respect to all images of a group. Although, at first sight, the shape of the histograms seems to be similar for the two groups, the histograms indicate that the mean segment length are reduced in the case of TGFα–treated cells compared with the untreated ones. Note that due to the applied image segmentation algorithm the minimal length of a line segment has to be 8 pixels (≈ 20 nm) which equals the merging parameter $d_{\max}$ chosen in Section 8.1.1. Thereby the mean number of detected segments is of course reduced. Figure 8.4 displays boxplots for the total number of segments and the mean segment length per sampling region for the TGFα–treated and the untreated group, respectively. Note that in this particular case it does not make a difference to regard total numbers of segments per sampling regions instead of mean numbers per unit area, since the sampling regions are of a fixed and equal size. Tests based on the Wilcoxon–Mann–Whitney test ([114]) show that the hypotheses of equality of the mean number of segments and equality of mean segment length for the two groups are both rejected at a significance level of 5 %.

Results of orientation analysis
In Figure 8.5 sample histograms representing the estimated distribution of the angles are displayed for both groups. Within each group a strong variability of the shape of the histograms can be noted. Nevertheless, it seems to be the case that the orientation distribution for the TGFα–treated group seems to be more uniform than the orientation distribution for the untreated group. Kuiper's test for uniform orientation distribution shows that for the untreated group seven of eight sample images lead to a rejection of the null–hypothesis of uniform orientation distribution for a test level of 5 %. For

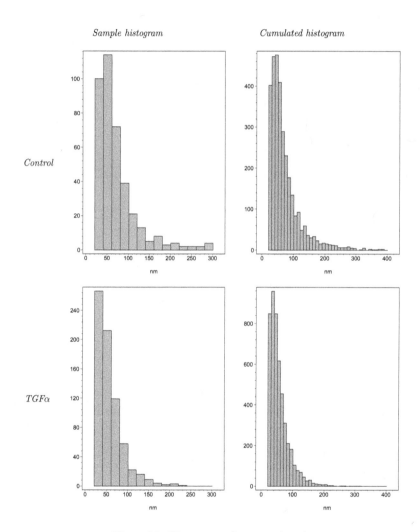

Figure 8.3: Histograms of segment length

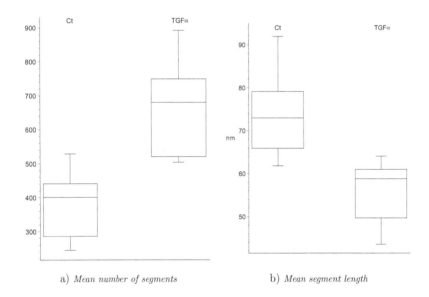

a) *Mean number of segments*   b) *Mean segment length*

Figure 8.4: Boxplots of the mean number of segments and mean segment length (the median is depicted as the line in the middle of the box, the first and third quartile as the upper and lower border of the box, and the minimal and maximal value are depicted as the upper and lower hook)

the TGFα–incubated group only four of seven sample images lead to a rejection of the same hypothesis at the same test level. To investigate this behaviour in more detail a boxplot of estimates for the circular standard deviation given in (8.7) for the two groups is shown in Figure 8.6. A Wilcoxon test for equality of expected circular standard deviations with respect to the two groups leads to rejection for a test level of 5 %. This means that the expected circular standard deviation is considered to be greater for the TGFα treated group than for the untreated group. Therefore, filaments of the control group seem to be more uniformly orientated than those of the TGFα treated group. In summary, together with the analysis of the segment lengths and segment numbers, the orientation analysis shows that there are distinct changes of the keratin network architecture in pancreatic cancer cells in response to TGFα. These changes will be investigated further in Section 8.1.3 on the basis of a fitting of suitable tessellation models.

8.1.3   Fitting of Nested Tessellation Models

In Section 8.1.2 it is shown that there are significant changes in the keratin network structure of pancreatic cancer cells caused by the incubation of TGFα. Whereas this investigation is based on a single–object analysis of the line segments we now turn our attention to an analysis of the global network architecture that can be achieved by an application of the model fitting procedure introduced in Chapter 7. As the results of this section will show, there is indeed a difference in the global architecture of the keratin network caused by the incubation with TGFα since different optimal tessellation models are detected for the two groups of untreated and TGFα–treated cells.

As a set of possible tessellation models for this example, one–fold nestings of random tessellations are chosen where the initial random tessellation models regarded comprise the Poisson line tessellation, the Poisson–Voronoi tessellation and the Poisson–Delaunay tessellation. With respect to the one–fold nesting the Bernoulli thinning probability p_B is chosen between 0.9 and 1 thereby reflecting the small probability of a non–iterated mesh, for example, due to vacuoles.

These three initial models and the nine possible combinations for one–fold nestings resulting from them in conjunction with the Bernoulli thinning probability already cover a broad spectrum of scenarios for cytoskeletal networks, although, regarded separately are still quite simple and theoretically tractable, for example, with respect to certain mean value formulae (cmp. Section 7.3). With respect to distance functions the relative Eulidean distance function is used, but results do not vary much if other (relative) distance functions are used.

Model fitting for single sample images
The results of the model fitting algorithm for single sample images are summarized in

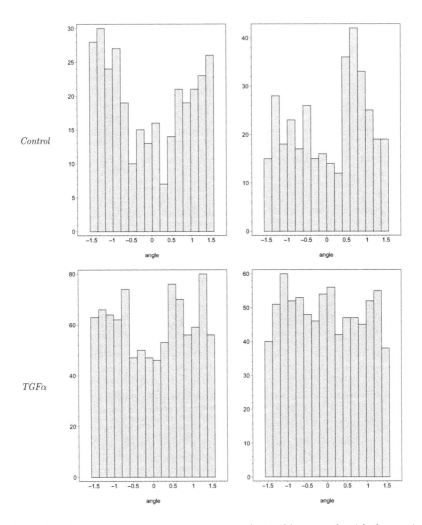

Figure 8.5: Histograms of segment orientations as depicted by its angle with the x–axis

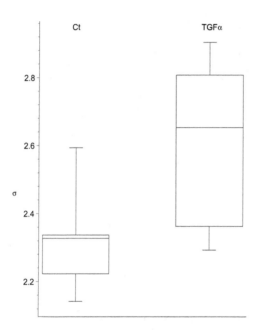

Figure 8.6: Boxplot of circular standard deviation (the median is depicted as the line in the middle of the box, the first and third quartile as the upper and lower border of the box, and the minimal and maximal value are depicted as the upper and lower hook)

Table 8.1: Optimal models for individual sample images of the group of untreated cells

sample	optimal model	distance
1	PVT/PLT	0.0325
2	PVT/PLT	0.0011
3	PVT/PLT or PLT/PLT	0.0019
4	PVT/PLT	0.0079
5	PVT/PLT or PLT/PVT	0.0161
6	PVT/PLT	0.0274
7	PVT/PLT	0.0071
8	PVT/PLT or PLT/PVT	0.0066

Tables 8.1 and 8.2. It can be deduced that for the group of untreated cells a relatively clear decision is made in favor of the PVT/PLT in all cases. This means that the PVT/PLT is always considered to be the optimal model with respect to the set of the nine possible models. Note that in some cases it is possible that the decision for an optimal model is not unique due to the symmetries in the corresponding formulae (cmp. Table 7.3), especially in the cases where for the Bernoulli thinning factor p_B it holds that $p_B = 1$. Looking at the group of cells that are treated with TGFα a different decision is viewable. While for four samples the optimal model is still a PVT/PLT, for the other three samples we have that the optimal model turns out to be a PDT/PLT. It is also important to note that for all samples we obtain that models for branched filaments (PVT or PDT) are preferred compared to an interaction–free model (PLT), thereby emphasizing that the keratin cytoskeleton is representing a branched filament network. Note that a comparison between the optimal nested model and the optimal basic model, i.e., the model that has minimal distance out of the three models PLT, PDT, PVT, yields a large improvement with respect to the quality of approximation (Tables 8.1 and 8.2) which can be deduced from the drastical reduction of the measured distances.

Model fitting for the vectors of mean characteristics
In order to enhance the results that are obtained for single sample images, we look at the results of the fitting procedure for the vectors of mean characteristics (cmp. Section 7.2.2). Recall that such a technique is possible since the samples (and therefore also the resulting vectors of mean values) can be considered to be independent and identically distributed. In Tables 8.3 and 8.4 the results of the fitting for the vectors

Table 8.2: Optimal models for individual sample images of the TGFα–incubated group of cells

sample	optimal model	distance
1	PDT/PLT or PLT/PDT	0.0104
2	PDT/PLT	0.0146
3	PDT/PLT or PLT/PDT	0.0138
4	PVT/PLT or PLT/PLT	0.0040
5	PVT/PLT	0.0044
6	PVT/PLT or PLT/PLT	0.0026
7	PVT/PLT or PLT/PVT	0.0050

of mean characteristics for the two groups are displayed. For the group of untreated cells the decision is in favor of a PVT/PLT as it has been the case for all individual sample images. For the group of TGFα treated cells the optimal model is a PDT/PLT, enhancing the effect observed for individual sample images that there is a structural difference between the filament network architecture of the two regarded groups.

Segmented graph structures of both groups, as a result of the applied image segmentation algorithm, together with sample realisations of the optimal models for the vectors of mean characteristics, are displayed in Figure 8.7. Clearly, the similarity between these illustrations can not be perfect, since there is a certain variability between different realisations of the same model.

8.1.4   Summary of the Results

The results of Sections 8.1.2 and 8.1.3 show that TGFα induces a profound change of the keratin filament network in pancreatic cancer cells, demonstrating that an approach based on the image segmentation techniques introduced in Chapter 6 and on the model fitting algorithm explained in Chapter 7 can provide important insights into the processes that govern network morphology in the cytoskeleton. In particular, with respect to the network structure, we can deduce from the results of Section 8.1.3 that there is a remodelling caused by TGFα. Whereas the untreated cells have uniformly the PVT/PLT model as a best fit, the cells after stimulation with TGFα show a mixed behaviour between PVT/PLT and PDT/PLT. A possible explanation for this behaviour is that some cells have not finished yet the transition and therefore still indicate PVT/PLT

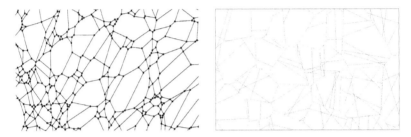

a) *Segmented graph structure and sample realisation of optimal PVT/PLT model for the group of untreated cells*

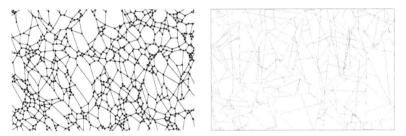

b) *Segmented graph structure and sample realisation of optimal PDT/PLT model for the group of cells stimulated with TGFα*

Figure 8.7: Comparison between segmented image data and realisations of optimal model for both groups

Table 8.3: Optimal parameter choices for possible models concerning mean values for the group of untreated cells

Model	Distance	γ_0	γ_1	p
One–fold nestings				
PLT/PLT	0.0057	0.022031	0.025125	0.9
PLT/PVT	0.0205	0.031608	0.000057	0.9
PLT/PDT	0.0560	0.029103	0.000033	0.9
PVT/PLT	**0.0015**	0.000049	0.033860	0.9
PVT/PVT	0.2023	0.000298	0.000030	1.0
PVT/PDT	0.1054	0.000101	0.000083	0.9
PDT/PLT	0.0447	0.000026	0.0331583	0.9
PDT/PVT	0.1123	0.000067	0.000123	0.9
PDT/PDT	0.1477	0.000053	0.000070	0.9
Basic tessellations				
PLT	0.2994	0.048714	-	-
PVT	0.2077	0.000539	-	-
PDT	0.8492	0.000362	-	-

Table 8.4: Optimal parameter choices for possible models concerning mean values for the group of cells stimulated with TGFα

Model	Distance	γ_0	γ_1	p
One–fold nestings				
PLT/PLT	0.0084	0.040824	0.018099	1.0
PLT/PVT	0.0024	0.044430	0.000054	1.0
PLT/PDT	0.0129	0.040115	0.000042	0.9
PVT/PLT	0.0024	0.044430	0.000054	1.0
PVT/PVT	0.2311	0.000835	0.000000	1.0
PVT/PDT	0.0911	0.000131	0.000177	0.9
PDT/PLT	**0.0005**	0.000038	0.041888	0.915
PDT/PVT	0.1030	0.000145	0.000160	0.9
PDT/PDT	0.1068	0.000087	0.000121	0.9
Basic tessellations				
PLT	0.2605	0.063116	-	-
PVT	0.2312	0.000827	-	-
PDT	0.8103	0.000622	-	-

as an optimal model, while others are already in a structural state that indicates an optimal PDT/PLT model. For more details on biological interpretations of these results see [8] and [9].

8.2   Model–based Analysis of Actin Filament Networks

The second example from cell biology where techniques from Chapters 6 and 7 are applied deals with the actin filament network in the cytoskeleton in lamellipodia. The actin filament network regulates the elasticity of whole cells and, consequently, influences cell migration ([11], [80]). It is formed by two distinct structures, lamellipodia and filopodia with the latter arising from the dendritic actin network of lamellipodia ([82], [102]). Although lamellipodia represent a very thin cytoplasmic compartment, they can contain several superimposed layers of flat actin networks ([103]). Therefore a topological problem arises that has to be dealt with when extracting structural properties from the (usually two dimensional) images. In this section we analyze a set of sample images showing the lamellipodia of the mouse melanoma cell line B16F1 with respect to these structural properties. In particular, after a description of image acquisition and image segmentation in Section 8.2.1, we apply in Section 8.2.2 the model fitting algorithm introduced in Chapter 7 in order to fit a one–fold superposed tessellation model to the given data, thereby representing a tessellation model for the actin network structure that consists of two independent layers. The results of this model fitting algorithm show that the actin filament network possesses two distinctive layers. Based on these layers it is possible to compute important characterstics for the structural properties of a cell as, for example, the elastic modulus which is done in Section 8.2.3. The results of Section 8.2 are also partially documented in [26].

8.2.1   Image Segmentation

Images were provided by T. Svitkina (Department of Biology, University of Pennsylvania) For details on the cell culture, specimen preparation and on image acquisition see [103]. Altogether, seven sample images from four different cells are analysed with varying sizes, where the total area is 10.91 μm^2 (length of 1 pixel $\approx 1.1\ nm$). Recall once more that, especially with respect to the accuracy of the analysis of vectors of mean characteristics, it is more important how large the total area of the investigated images is compared to the total number of images. The sample images are smaller cutouts of larger images, where the sample images are chosen such that they have a varying distance to the cell membrane and that artifacts caused by preparation and imaging are mostly avoided (Figure 8.8).

Due to the constant diameter of actin filaments in the images, noise effects can be reduced by applying a Gaussian filtering (Section 6.1.2) with a fixed filter size of 10 pixels which reflects quite well the width of the filaments. A watershed algorithm based on immersion (Section 6.3) is applied to the filtered image in order to detect the actin cytoskeleton that can be approximated by a network of line segments where all endpoints are connected, i.e., where there are no dead ends. The pixels belonging to the resulting dam structure are classified into crosspoints and linepoints (Section 6.4.1). Afterwards, the dam structure is compared to the original image. With respect to each dam we regard the grey scale value at the corresponding pixel coordinates. In case that the minimum of these values is bigger than a given threshold, the dam is kept, otherwise it is deleted from the structure. Such a procedure is used in order to remove dams that result from topological disturbances and that do not correspond to actin filaments. The dams are replaced by straight lines between the two endpoints of the dam assuming that a straight line approximates an actin filament due to its sufficiently large persistence length. Finally, we correct topological disturbances of the primary segmentation by merging neighboured branching points (Section 6.4.3), where the merging parameter $d_{\max}$ is chosen as 10 pixels reflecting the width of the filaments. This means that branching points are merged in an ascending order if they are less than $d_{\max} = 10$ pixels apart from each other. Results of the image segmentation algorithm for an increasing complexity of the original image are displayed in Figure 8.9.

8.2.2 Fitting of Superposed Tessellation Models

In order to obtain objective features of global network architecture random tessellation models are fitted to the graphs resulting from the segmentation of the original images showing the actin filament networks using the methods described in Chapter 7. Assuming that the actin network consists of at least two independent two–dimensional layers a one–fold superposition is fitted to the data. Basic tessellation models for this superposition are given by PLT, PVT, and PDT which results in six different possible combinations for the one–fold superposition (Figure 7.5). As a distance function the relative Eulidean distance is applied.

The results of the model fitting for the vector of mean characteristics are shown in Table 8.5. From these results it can be deduced that the optimal one–fold superposed model is given by a PVT/PDT superposition, where the two parameters are given by $\gamma_{PVT} = 0.000582$ and $\gamma_{PDT} = 0.000041$ (recall that the parameters γ have a different meaning for different basic tessellation types, cmp. Sections 2.4.3 –2.4.5). The results for the vector of mean characteristics are validated by the results for the optimal model of single sample images displayed in Table 8.6. Here, in each case a PVT/PDT superposition is the optimal model, always with parameters in a narrow range. Recall the fact that a PVT/PDT superposition is equivalent with respect to distribution to a PDT/PVT superposition. By a comparison of non–superposed to one–fold superposed models it becomes visible that there is a large improvement in the quality of approx-

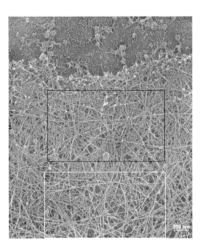

a) *Overview of the actin filament network*

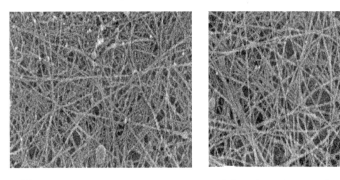

b) *First sample image of actin network*   c) *Second sample image of actin network*

Figure 8.8: Actin filament networks in B16F1 mouse melanoma cells

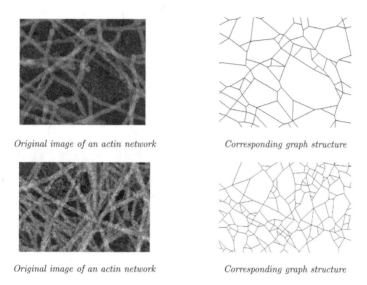

Original image of an actin network   *Corresponding graph structure*

Original image of an actin network   *Corresponding graph structure*

Figure 8.9: Results of image segmentation for actin networks

imation that manifests itself in a remarkably reduced measured distance. A sample image and a realisation of the optimal superposed model are given in Figure 8.10.

8.2.3 Estimation of Actin Network Elasticity

A characteristic that is of vital interest with respect to actin networks is the elastic shear modulus G. It describes the tendency of the network to shear that means the deformation of shape at constant volume, when acted upon by opposing forces. The elastic shear modulus is defined as shear stress over shear strain. In order to determine G for the actin networks analysed in Sections 8.2.1 and 8.2.2 an approach will be introduced that is based on the model fitting performed and on the determination of the mean mesh size for the optimal superposed tessellation model. The mean mesh size is determined since the value for the elastic shear modulus is directly dependent from it and approximately scales as mean mesh size to the power of -4.4. Note that in general we can derive an estimation for the mean mesh size only via simulations, especially with regard to one–fold superposed tessellation models as used in the following. As a control computation also the measurement of the actin concentration that can be derived from

Table 8.5: Goodness–of–fit and model parameters for actin networks

Model	Distance	γ_0	γ_1
Basic tessellations			
PLT	0.2496	$6.981 * 10^{-2}$	-
PVT	0.2534	$1.086 * 10^{-3}$	-
PDT	0.7369	$7.905 * 10^{-4}$	-
One–fold superpositions			
PLT/PLT	0.2496	$3.491 * 10^{-2}$	$3.491 * 10^{-2}$
PLT/PVT	0.0562	$2.167 * 10^{-2}$	$5.488 * 10^{-4}$
PLT/PDT	0.2467	$6.850 * 10^{-2}$	$1.168 * 10^{-7}$
PVT/PVT	0.0584	$2.920 * 10^{-4}$	$2.920 * 10^{-4}$
PVT/PDT	**0.0322**	$5.802 * 10^{-4}$	$4.116 * 10^{-5}$
PDT/PDT	0.3923	$1.254 * 10^{-4}$	$1.254 * 10^{-4}$

Table 8.6: Optimal models for individual sample images of the actin network

sample	optimal model	distance
1	PVT/PDT	0.0140
2	PVT/PDT	0.0026
3	PVT/PDT	0.0109
4	PVT/PDT	0.0383
5	PVT/PDT	0.0404
6	PVT/PDT	0.0273
7	PVT/PDT	0.0698

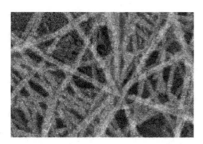 

Figure 8.10: Comparison between sample image and realisation of optimal PVT/PDT superposed model fitted to actin networks

the total length of the segmented filament structures is used. Thereby an alternative estimate for the average mesh size is obtained that can be used to estimate G.

With respect to an isotropic crosslinked actin filament network we can calculate G in general as ([51])

$$G = 6k_b T l_p^2 / (l_e^3 \zeta^2) \tag{8.8}$$

with k_b being the Boltzman constant ($1.3806505 * 10^{-23} \ J/K$), T the temperature in K, l_p being the persistence length of actin filaments ($17 \ \mu m$), l_e the entanglement length and ζ the average mesh size. The entanglement length l_e is defined as the average distance between points along an actin filament that are effectively constraint. For a fully crosslinked network (i.e., a network where all crossings are really linked and not only overlapping) l_e is determined by the distance between the crosslinks. In [43] it is stated that the entanglement length is $2.2 \ \mu m$ for an actin network with a concentration of $1 \ mg/ml$ and that it is proportional to $\rho^{-0.4}$. The average mesh size ζ represents the average distance between neighbouring actin filaments. Due to the fact that we assume a superposition of filament layers, ζ can not be determined directly from the observed images but has to be estimated from the optimal tessellation models. For this purpose we estimated the average mesh size ζ by estimating the average maximum diameter of circles inscribed into the meshes (i.e., cells) of simulated networks for each of the two layers of the optimal model (PVT and PDT) seperately. This leads to estimations for ζ of $34 \ nm$ with respect to the PVT layer and to $78 \ nm$ with respect to the PDT layer. Note that the estimations have to be performed by simulation since no analytical formulae are available.

As a control (or alternative) computation for the average mesh sizes that we obtained by simulations of the fitted one–fold superposed model we used the following. In [73] it is stated that the mesh size of an actin network depends on its actin concentration ρ

and that this relation can be described by

$$\rho = 1/\zeta^2. \tag{8.9}$$

So, as a first step in order to determine the concentration of filamentous actin, we calculated the total amount of filamentous actin per unit area by measuring the total length of the skeletal lines in the images and dividing by the total area regarded afterwards. Note that the precision of this method is hardly affected by a possible overlay of actin filaments from different layers since the skeletal lines have a negligible width and since they are mainly oriented orthogonally to the vertical axis. With respect to the total length of the skeletal lines per unit area we obtained $7.24 * 10^{-2}$ $nm/nm^2 =$ $7.24 * 10^1$ $\mu m/\mu m^2$. Using the optimal parameter values obtained for the two different optimal layers ($\gamma_{PVT} = 0.000582$ and $\gamma_{PDT} = 0.000041$) obtained in Section 8.2.2 we get the estimation that approximately 68.8% which is equivalent to $4.98 * 10^1$ $\mu m/\mu m^2$ belong to the PVT layer, whereas 31.2% ($2.26 * 10^1$ $\mu m/\mu m^2$) belong to the PDT layer. Due to the thickness of the lamellipodia which can be assumed as approximately 200 nm and if we additionally assume that the two layers are both occupying the same amount of height (100 nm each), we are able to rewrite the two total lengths as total lengths per unit volume and obtain $4.98 * 10^2$ $\mu m/\mu m^3$ for the PVT layer and $2.26 * 10^2$ $\mu m/\mu m^3$ for the PDT layer. Furthermore, by assuming that an actin filament with a length of 1 μm consists of 370 molecules with a molecular weight of 43 kDa per molecule (cmp. [104]) we finally obtain actin concentrations of 13.25 mg/ml for the PVT layer and 5.95 mg/ml for the PDT layer. Plugging these values into (8.9) average mesh sizes ζ of 44 nm and 66 nm for the PVT layer and the PDT layer respectively are obtained that conincide quite well with the estimates obtained from the simulated optimal tessellation models (34 nm and 78 nm).

Based on (8.8) and by using the estimates for ζ with respect to the two tessellation layers (34 nm for PVT and 78 nm for PDT) we are now able to derive estimations for the elastic shear modulus G as 23.4 kPa (PVT layer) and 0.6 kPa (PDT layer), respectively. Note that (8.8) refers to isotropic 3D networks and although the actin networks in lamellipodia basically represent 2D structures, this formula provides good estimates for G.

8.2.4   Summary of the Results

Due to the observation that lamellipodia are composed of flat actin filament networks ([97]), in Section 8.2.2 a suitable one–fold superposed tessellation model has been fitted to the actin filament networks. The usage of statistical tessellation models allows the coverage of a wide range of morphological scenarios without restricting to specific topological patterns. In particular the fact that the optimal model for all sample images is given by a PVT/PDT superposed tessellation with a very narrow range of optimal parameters shows that this model is quite convenient for modelling the observed network structures and that a uniform network morphology exists for this cytoskeletal compartment. These impression are further enhanced by visual comparisons. Note that the modelling of actin filament networks as a one-fold superposition fits well with previous observations ([83]), which have shown that the lamellipodium consists of two distinct actin networks.

In Section 8.2.3 the fitted model has been used in order to obtain estimates for the mean meshwidths of the two layers (PVT and the PDT), which then have been utilised to derive estimations for the elastic shear modulus G by estimating the average mesh sizes. An alternative way in order to estimate G is by using the actin concentration that can be estimated by measuring the mean total length of actin filaments per unit area. This total actin concentration in lamellipodia was measured to be 9.6 mg/ml which is in accordance to previously measured data ([116]). The calculation of the total filament length by measuring the projection of filaments provides good estimates of actin concentrations since actin networks appear to be spatially restricted ([97]). The superposed tessellation model fitted to the data provided the opportunity to analyse the structural properties of the different actin network layers separately, where this analysis revealed that the two layers are non-identical and differ with respect to morphology and actin concentration. By a comparison of the density of actin filaments determined for the different layers and the original images it can be deduced that the lower layer of the actin network can be described by the PVT layer, while the upper layer is represented by the PDT model. The comparison of the estimates for the average mesh sizes of 34 nm for the PVT layer and of 78 nm for the PDT layer, respectively, to an estimate obtained purely by considering the actin concentrations (44 nm and 66 nm for the PVT layer and the PDT layer, respectively) show that they are of comparable sizes, differing only by 18% and 23%, respectively, from those calculated from actin concentrations ([73]). This variation might be caused by an unequal division of the cytoplasmic space between the two layers of the actin network which affects the concentration values directly. However, these small differences show that the properties of the fitted model indeed reflect the structural characteristics of the multi–layer actin network. Based on the estimations for the mesh sizes, the elastic shear modulus G was found to be 23.4 kPa for the PVT and 0.6 kPa for the PDT layer, respectively, which is within the range of previously measured data ([3]). The difference in the value of G for the two different layers may cause an asymmetry of the elastic properties of the lamellipodium, where the stiffer layer determines the leading edges ability to push the cell forward. The response to small forces perpendicular to the lamellipodium however might be regulated by the softer PDT layer. For a more detailled discussion of the biological aspects for the results of Section 8.2 the reader is referred to [26].

Chapter 9

Conclusions and Outlook

This thesis has shown that techniques and tools from stochastic geometry and image analysis can be applied to different fields of applications like cell biology and telecommunication almost independent of the scale the corresponding data is measured in. In particular efficient algorithms for the simulation of the typical cell for different types of tessellations have been developed and described. Characteristics for the typical cell like the area or the perimeter have been analysed which can provide information in different kinds of applications like transport problems in membrane cell trafficking or cost analysis in telecommunication networks. Examples for such a cost analysis based on the simulation algorithms for the typical cell have been performed and inference for different cost characteristics like the mean shortest path length or the mean distance to the nearest cell nuclei have been obtained. Problems connected to the correct and efficient implementations of the corresponding algorithms either for typical cell simulation or for cost analysis have also been discussed. For two sets of images from cell biology, network structures have been extracted by means of morphological analysis. These network structures have been analysed with respect to their basic statistical characteristics like the mean number of segments per image or the orientation distribution of the segments. Additionally suitable iterated tessellation models have been fitted to these networks using an automatised fitting procedure. These fitted random tessellation models have afterwards been used in order to obtain valuable information about geometric features of the networks like the average mesh size and the elastic shear modulus. Obviously, such an approach is not limited to examples from cell biology but has already been successfully applied to an example coming from telecommunication (cmp. [31]).

Looking at the results of this thesis the question for extensions and further development naturally arises. For example, with respect to the development of algorithms for the simulation of the typical cell different types of tessellations or partitions based on

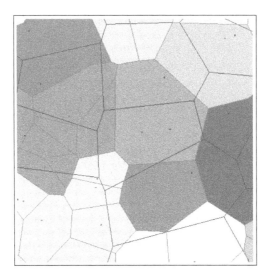

Figure 9.1: Realisation of an aggregated Poisson–Voronoi tessellation, where different colours denote different cells

tessellations can be considered. In cooperation with M. Wauer some preliminary results have been obtained for the simulation of the typical cell for aggregated Poisson–Voronoi tessellations (cmp. [5], [29], [105]) which are partially documented in his diploma thesis ([112]). A realisation of an aggregated Poisson–Voronoi tessellation is displayed in Figure 9.1. In particular, a comparison of the results for the simulation algorithm with respect to the distribution of the area for the typical cell to an approximation formula based on known distributions for tessellations in the Poisson–Voronoi case has been performed. With respect to tessellations that are based on modulated Poisson point processes, algorithms for the simulation of the typical cell for modulated Delaunay tessellations as well as for multi–modulated tessellations are thinkable. Multi–modulated in this context means that instead of one Boolean model that induces the random driving measure for the modulated Poisson point process we have several Boolean models with a certain hierarchy among them and different intensities connected to them.

With regard to cost analysis a natural extension of the results shown in this thesis is to regard distributional properties of the cost characteristics apart from the first moment. In cooperation with D. Wolfmüller some preliminary results have been obtained for the estimation of the densities for the shortest path length as well as for the subscriber

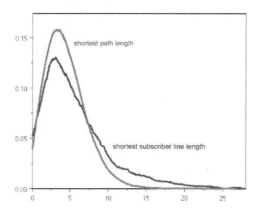

Figure 9.2: Estimated densities of shortest path length and subscriber line length for $\kappa = 50$ (taken from [115])

line length (cmp. Figure 9.2 and Section 4.2) which are documented in his diploma thesis ([115]). Other cost functionals apart from the Euclidean distance for the costs on a single segment have also been regarded in this diploma thesis, for example, an assignment of a constant cost to a line segment that is independent of the segment length. Such a choice of the costs leads to problems of capacity analysis which will also be interesting to analyse in the future.

A different aspect is to change the underlying tessellation model for the cost analysis. In this thesis we have regarded Voronoi tessellations that are based on a Cox point process that is concentrated on the lines of a Poisson line process. It might instead be interesting, for example, to regard models that are based on a Cox point process that is concentrated on the edges of a Poisson–Voronoi tessellation or on the edges of a one–fold nesting of tessellations (cmp. Figure 9.3). Especially by a comparison to the case regarded in this thesis interesting information about the influence of the underlying geometrical structure on the cost functionals can be obtained.

The model fitting algorithm introduced in Chapter 7 can be refined and extended. For example, more input parameters for the vector of input characteristics can be considered. Possible candidates are the angles between segments or the number of neighbouring vertices with respect to a vertex. Of course it might be difficult to derive theoretical formulae for mean values of these characteristics and hence the computation has to be done by simulation. Another interesting topic is the generation of intensity maps or even type–intensity maps from the model fitting. This means that based on

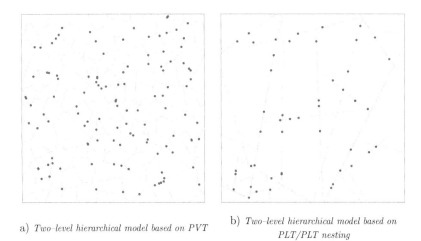

a) *Two–level hierarchical model based on PVT*

b) *Two–level hierarchical model based on PLT/PLT nesting*

Figure 9.3: Different underlying geometrical models for two–level hierarchies

the algorithm for different locations as center point of the sampling window an optimal model together with corresponding parameters is determined. If this is done for a single tessellation model thereby intensity maps can be generated using an extrapolation technique like kriging (cmp. [111]). If different optimal random tessellation models are obtained, this method might be extended to type–intensity maps, where various possible approaches of how to handle borders between regions of different optimal models are thinkable. First results for intensity maps with respect to urban infrastructure data of Paris are documented in [106]. A similar approach might be helpful for the investigation of the filamentous networks that have been investigated in this thesis. In this context of course a problem of image acquisition occurs since the size of the images is restricted by technical constraints of the electron microscope.

Another interesting aspect for future work is the introduction of certain dynamics to the scenarios investigated in this thesis. For example, the location of the lower–level points in the two–level hierarchical models for cost analysis might no longer be fixed but they might be moving according to a dynamic model. This leads to models that might be able to reflect mobile telecommunication settings quite well. Also for the cases regarded in cell biology dynamics are worthy to add to the model. Here an interesting question is how to model the transition from one network state into another one like after injection of TGFα into the keratin filament network. Some preliminary results of such a modelling are summarised in [50].

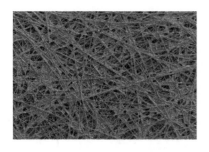
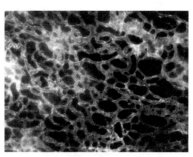

a) *Gas–diffusion layer in a fuel cell*   b) *Blood arteries in a mouse*

Figure 9.4: Network structures occurring in material science and medicine

A challenging problem for the future will be the extension of the techniques to three–dimensional data. With the progress in imaging techniques like 3D–electron tomography or similar methods more and more 3D data will be available soon. While most of the techniques introduced here are extendable to the third dimension of course some adaptions have to be performed. For example, with regard to image segmentation using more sophisticated algorithms like grey scale skeletonization ([109]) have to be used.

Finally we would like to mention that as it has been stated in the beginning all the techniques described in this thesis are not restricted to the fields of telecommunication and cell biology but can also be applied to various other fields where similar network structures are occurring, for example, in material science and medicine (cmp. Figure 9.4, graciously provided by the ZSW, Ulm and by the Institute of Physiological Chemistry, Ulm University, respectively).

Appendix A

Basic Mathematical Definitions

In the appendix a short overview of some basic definitions and theorems in set theory, topology, measure theory, and probability calculus is provided. For a more detailed description of these topics the reader is referred to the literature, for example, to [7], [13], [19], [45], and [89].

A.1   Set Theory

We describe by a set M a collection of arbitrary distinct objects, where a single object x as part of this set is called an *element* and is denoted by $x \in M$. The notation $x \notin M$ denotes an object x that does not belong to M. A set A is said to be a *subset* $A \subset M$ if for all $x \in A$ we have that $x \in M$. A set that contains no elements is called an *empty set* and is denoted by $\emptyset$. As customary, the *union* and the *intersection* of two sets $A, B \subset M$ can be defined as

$$A \cap B = \{x : x \in A \text{ and } x \in B\}$$

and

$$A \cup B = \{x : x \in A \text{ or } x \in B\},$$

respectively. Furthermore, let A^c denote the *complement* of A defined by

$$A^c = \{x : x \in M \text{ and } x \notin A\},$$

and let

$$A \setminus B = \{x : x \in A \text{ and } x \notin B\}$$

be the *difference* $A \backslash B$. For arbitrary sets $M_1, ..., M_n$ we can define the *Cartesian product* as the set of all vectors $(x_1, ..., x_n)$, where $x_i \in M_i$. It is denoted by $M_1 \times ... \times M_n$ with

$$M_1 \times ... \times M_n = \{(x_1, ..., x_n) : x_1 \in M_1, ..., x_n \in M_n\}.$$

In this thesis mainly the two–dimensional set of real numbers $\mathbb{R}^2$ is regarded. Therefore we now recall some operations on $\mathbb{R}^2$.

The elements of $\mathbb{R}^2$ are called *points* or *vectors*. A point $x \in \mathbb{R}^2$ consists of 2 real–valued components, in other words $x = (x_1, x_2)$, where $x_i \in \mathbb{R}$ for $i = 1, 2$. The point $o = (0, 0) \in \mathbb{R}^2$ is referred to as the origin. In addition, a *scalar multiplication* of $x \in \mathbb{R}^2$ by a number $c \in \mathbb{R}$ is given by

$$cx = (cx_1, cx_2).$$

The *addition* of two vectors $x, y \in \mathbb{R}^d$ is defined in a componentwise fashion as

$$x + y = (x_1 + y_1, x_2 + y_2).$$

For $A \in \mathbb{R}^2$ and $c \in \mathbb{R}$ let

$$cA = \{cx : x \in A\}.$$

In particular, if $(-1)A = A$ the set $A \subset \mathbb{R}^2$ is called *symmetric*. The *translation* of a set $A \subset \mathbb{R}^2$ by a vector $x \in \mathbb{R}^2$ is given by

$$A_x = A + x = \{y + x : y \in A\},$$

while the rotation of the set $A \subset \mathbb{R}^2$ around the origin can be defined as

$$\vartheta_R(A) = \{Rx : x \in A\},$$

where $R \in \mathbb{R}^{2 \times 2}$ is an orthogonal matrix with $\det R = 1$. The *Minkowski sum* $A \oplus B$ of two sets $A, B \in \mathbb{R}^d$ is given as

$$A \oplus B = \{x + y : x \in A, y \in B\}.$$

It can easily be shown that the operation '$\oplus$' is associative as well as commutative.

A.2   Topology

Let $M \subset \mathbb{R}^2$ be an arbitrary set with elements $x, y \in M$. A mapping $\rho : M \times M \to \mathbb{R}_+$ is called a *metric* with respect to $M \subset \mathbb{R}^2$ if the following properties hold

1. $\rho(x, y) = 0$ if and only if $x = y$,

2. $\rho(x, y) = \rho(y, x)$, and

3. $\rho(x, z) \leq \rho(x, y) + \rho(y, z)$, $\forall\, x, y, z \in M$.

The tuple (M, ρ) then denotes a *metric space*.

A sequence $x_1, x_2, \ldots$ in (M, ρ) is called a *Cauchy sequence*, if for any $\epsilon > 0$ there exists a number $n_0 \in \mathbb{N}$ such that for $n, n' > n_0$ it holds that

$$\rho(x_n, x_{n'}) < \epsilon.$$

A metric space (M, ρ) is called *complete*, if all Cauchy sequences in (M, ρ) are converging. This means that for any Cauchy sequence $x_1, x_2, \ldots$ in (M, ρ) there exists a limit $x \in M$ such that for any $\epsilon > 0$ there is a number $n_0 \in \mathbb{N}$ with

$$\rho(x_n, x) < \epsilon$$

for all $n \geq n_0$.

Let (M_1, ρ_1) and (M_2, ρ_2) be two metric spaces. A mapping $f : M_1 \to M_2$ is called *continuous* in $x_0 \in M_1$ if for all $\epsilon > 0$ there exists a $\delta(\epsilon, x_0)$ such that $\rho_2(f(x_0), f(x)) < \delta(\epsilon, x_0)$ holds for any x with $\rho_1(x_0, x) < \epsilon$.

The *closure* $cl\, A$ of a set $A \subset M \subset \mathbb{R}^2$ is defined as the intersection of all open subsets of M that contain A. The subset A is called *dense* in M if $cl\, A = M$ and the set M is called *separable* if it contains a countable, dense subset.

Consider a metric space (M, ρ) and let $b(a, r) = \{x \in M : \rho(x, a) \leq r\}$ be the ball with center $a \in M$ and radius r. Then a subset $A \subset M$ is called

- *bounded* if there is an $a \in \mathbb{R}^d$ and an $r > 0$ such that $A \subset b(a, r)$,

- *open* if $\forall\, x \in A\ \exists\, \epsilon > 0 : b(x, \epsilon) \subset A$,

- *closed* if its complement $A^c = M \setminus A$ is open,

- *convex* if $\lambda x + (1 - \lambda)y \in A$ for arbitrary $x, y \in A$ and $\lambda \in (0, 1)$.

Let E be a non–empty set and $\mathcal{T}$ a system of subsets of E. Then $\mathcal{T}$ is called a *topology* on E if the following holds

1. $\emptyset \in \mathcal{T}$, $E \in \mathcal{T}$,

2. if I is an index set and $O_i \in \mathcal{T}$ for all $i \in I$ then $\bigcup_{i \in I} O_i \in \mathcal{T}$,

3. if $O_1, ..., O_k \in \mathcal{T}$ then $\bigcap_{i=1}^{k} O_i \in \mathcal{T}$.

The tupel $(E, \mathcal{T})$ is then called a *topological space* and every element of $\mathcal{T}$ is called an *open set* of the topological space $(E, \mathcal{T})$.

Note that every metric space (E, ρ) is a topological space $(E, \mathcal{T})$, where $\mathcal{T}$ is the system of open sets in (E, ρ). The topology $\mathcal{T}$ is said to be *induced by* ρ. Analogously to a metric space, a topological space in which a countable, dense subset exists is called *separable*.

A system $\mathcal{B}_N = \{B_\nu | \ \nu \in N, N \text{ index set}\}$ of open subsets B_ν of the topological space $(E, \mathcal{T})$ is a *basis* of $\mathcal{T}$ if any open set in $(E, \mathcal{T})$ is a union of sets from $\mathcal{B}_N$. A topological space is called a *Polish space* if there exists a complete metric that defines the topology and if the topology possesses a countable basis.

Maybe the most prominent example of a metric space is $\mathbb{R}^2$ equipped with the *Euclidean metric* $|.|$ that is defined as

$$|x - y| = \sqrt{(x_1 - y_1)^2 + (x_2 - y_2)^2},$$

for all $x = (x_1, x_2)$ and $y = (y_1, y_2)$ in $\mathbb{R}^2$. Due to the fact that the set $\mathbb{Q}^2$, the set of all points in $\mathbb{R}^2$ whose coordinates are rational, is countable and dense in $\mathbb{R}^2$ we obtain that $(\mathbb{R}^2, |\cdot|)$, the metric space $\mathbb{R}^2$ equiped with the Euclidean metric is a complete, separable, metric space. A subset $A \subset \mathbb{R}^2$ is called *compact* if it is closed and bounded.

A.3   Measure Theory

Let $\Omega \neq \emptyset$ be an arbitrary set and let $\mathcal{P}(\Omega)$ be the power set of Ω. A system of sets $\mathcal{A} \subset \mathcal{P}(\Omega)$ is called a σ–*algebra* with respect to Ω if

1. $\Omega \in \mathcal{A}$,

2. $A \in \mathcal{A} \Longrightarrow A^c \in \mathcal{A}$, and

3. $A_i \in \mathcal{A}, i \in \mathbb{N} \Longrightarrow \bigcup_{i=1}^{\infty} A_i \in \mathcal{A}$.

A tupel $(\Omega, \mathcal{A})$, where $\mathcal{A}$ is a σ–algebra with respect to Ω is called a *measurable space*. A prominent example for a σ–algebra is the *Borel–σ–algebra* denoted by $\mathcal{B}(\Omega)$. It is defined as the smallest σ–algebra with respect to the set Ω which contains all open subsets of Ω. The corresponding measurable space is denoted by $(\Omega, \mathcal{B}(\Omega))$. Note that in this example Ω together with some topology $\mathcal{T}$ has to be a topological space. In general, if E denotes a non–empty set and if $\mathcal{E}$ denotes a family of non–empty subsets of E, we call the set $\sigma(\mathcal{E})$ which is the smallest σ–algebra which contains $\mathcal{E}$ the σ–algebra that is induced by $\mathcal{E}$.

Let $(\Omega, \mathcal{A})$ be a measurable space. A mapping $\mu : \mathcal{A} \to \mathbb{R} \cup \{\infty\}$ is called a *measure* on the measurable space $(\Omega, \mathcal{A})$ if

1. $\mu(A) \geq 0, \forall A \in \mathcal{A}$,

2. $\mu(\emptyset) = 0$, and

3. $\mu\left(\bigcup_{i=1}^{\infty} A_i\right) = \sum_{i=1}^{\infty} \mu(A_i)$ for any $A_i \in \mathcal{A}$, $i \in \mathbb{N}$ with $A_k \cap A_l = \emptyset$ for any $k, l \in \mathbb{N}$ with $k \neq l$,

where the third property is called the σ–*additivity* of the measure μ. If the mapping μ is of the form $\mu : \mathcal{B}(\Omega) \to [0, \infty]$, the mapping μ is called a *Borel measure*. A measure $\mathbb{P}$ is called a *probability measure* if $\mathbb{P}(\Omega) = 1$. A triple $(\Omega, \mathcal{A}, \mu)$ that is constructed by a measurable space $(\Omega, \mathcal{A})$ and a measure μ is a *measure space*. In the case that the measure is additionally a probability measure, the measure space is referred to as a *probability space*. Measures with non–negative integer values are called *counting measures* and a measure μ on a measurable space $(\mathbb{R}^2, \mathcal{B}(\mathbb{R}^2))$ is said to be *locally finite* or a *Radon measure* if for all bounded subsets $B \in \mathcal{B}_0(\mathbb{R}^2)$ we have that $\mu(B) < \infty$. In case that the counting measure μ is defined with respect to a measurable space $(\Omega, \mathcal{A})$ and that furthermore $\{\omega\} \in \mathcal{A}$ for all $\omega \in \Omega$ we can define the *support* of μ as the set supp $\mu = \{\omega \in \Omega : \mu(\{\omega\}) > 0\}$ A counting measure μ is called *simple* if $\mu(\{\omega\}) = 1$ for all $\omega \in$ supp μ almost surely.

The *Dirac measure* is an example of a simple counting measure. It is of the form $\delta_\omega(A) = 1$ if $\omega \in A$ and 0 otherwise. Another very prominent measure is the (two–dimensional) *Lebesgue measure* ν_2 on the two–dimensional measurable space $(\mathbb{R}^2, \mathcal{B}(\mathbb{R}^2))$. The Lebesgue measure is uniquely defined by its property that it assigns to each half–open rectangle $B \in \mathcal{B}(\mathbb{R}^2)$ with $B = [a_1, b_1) \times [a_2, b_2)$ its corresponding area $\nu_2(B) = (b_1 - a_1) \cdot (b_2 - a_2)$. For the Lebesgue measure we have the following properties

- ν_2 is *motion–invariant*, i.e. $\nu_2(B) = \nu_2(B_x) = \nu_2(\vartheta(B))$ for any $B \in \mathcal{B}(\mathbb{R}^2)$, where B_x is the set B shifted by an arbitrary vector $x \in \mathbb{R}^2$ and where $\vartheta(B)$ is the set B rotated around the origin by an arbitrary rotation ϑ;

- ν_2 is *locally finite*, i.e. $\nu_2(B) < \infty$ for all bounded sets $B \subseteq \mathbb{R}^2$, and

- ν_2 is a *diffuse* measure, i. e. $\nu_2(\{x\}) = 0$ for all $x \in \mathbb{R}^2$.

An important property of the Lebesgue measure is summarised in Haar's Lemma.

Lemma A.1 *(Haar's Lemma) Let $\mu : \mathcal{B}(\mathbb{R}^2) \to [0, \infty]$ be a locally finite and motion–invariant measure. Then there exists a constant $c \in [0, \infty)$ such that*

$$\mu(B) = c\nu_2(B)$$

for all $B \in \mathcal{B}(\mathbb{R}^2)$.

A proof of this lemma can be found, for example, in [36] on pp. 251f.

A function $f : \Omega \to \Omega'$, where $(\Omega, \mathcal{A})$ and $(\Omega', \mathcal{A}')$ are two measurable spaces, is called $(\mathcal{A}, \mathcal{A}')$–*measurable* if $f^{-1}(A') = \{\omega \in \Omega : f(\omega) \in A'\} \in \mathcal{A}$ for all $A' \in \mathcal{A}'$. If additionally $(\Omega', \mathcal{A}') = (\mathbb{R}^d, \mathcal{B}(\mathbb{R}^d))$ we call the function f *Borel–measurable*.

A function $f : \Omega \to (-\infty, \infty)$ is called μ–*integrable* if f is measurable and if the integrals $\int f^+ d\mu$ and $\int f^- d\mu$ exist, where $f^+(\omega) = \max\{0, f(\omega)\}$ and $f^-(\omega) = \max\{0, -f(\omega)\}$. In such a case we denote by

$$\int f d\mu = \int f^+ d\mu - \int f^- d\mu$$

the μ-*integral* of f with respect to Ω.

Let μ be a measure on a σ–algebra $\mathcal{A}$ of Ω. We call a set $N_0 \in \mathcal{A}$ a *null set* with respect to μ if $\mu(N_0) = 0$. Furthermore we say that a property η holds *(μ–)almost everywhere* or *(μ–)almost surely* on Ω if there exists a null set N_0 with respect to μ such that η holds for all $\omega \in \Omega \setminus N_0$.

A.4   Probability Calculus

Consider a probability space $(\Omega, \mathcal{A}, \mathbb{P})$ and a measurable space $(\Omega', \mathcal{A}')$. A *random variable* X is a measurable mapping $X : \Omega \to \Omega'$. In particular, we call a random variable $X : \Omega \to \mathbb{R}$ a *real–valued random variable*. Analogously a vector $X = (X_1, ...X_d)^\top$ of real–valued random variables $X_1, ...X_d$ is called a *random vector*. The *distribution* P_X of a random variable X is given by

$$P_X(A') = \mathbb{P}(X \in A') = \mathbb{P}(\{\omega \in \Omega : X(\omega) \in A'\}), \ A' \in \mathcal{A}'.$$

More specifically, if we consider a random vector $X = (X_1, ...X_d)^\top$ the function $F_X : \mathbb{R}^d \to [0, 1]$ with

$$F_X(x_1, ..., x_d) = \mathbb{P}(X_1 \leq x_1, ..., X_d \leq x_d) = P_X(X_1 \in (-\infty, x_1], ..., X_d \in (-\infty, x_d])$$

is called the *(cumulative) distribution function* (cdf) of X. With respect to the cdf a function that is often of interest is the quantile function. In general, if $F : \mathbb{R} \to \mathbb{R}$ is a non–decreasing and right–continuous function we have that the function $F^{-1} : \mathbb{R} \to \mathbb{R}$ given by

$$F^{-1}(y) = \inf\{x : F(x) \geq y\}$$

is called the *generalized inverse function* of F, where $\inf \emptyset = \infty$. Now, if we consider the cdf F_X of a random variable X on $(\Omega, \mathcal{A}, \mathbb{P})$ we call the function F_X^{-1} the *quantile function* of X. We say that a sequence $\{X_n\}_{n \geq 1}$ of real–valued random variables *converges in distribution* to a real–valued random variable X and write $X_n \xrightarrow{d} X$ if $F_{X_n}(x) \to F_X(x)$ for every point x such that $\mathbb{P}(X = x) = 0$.

If two random variables X_1 and X_2 on $(\Omega, \mathcal{A}, \mathbb{P})$ are considered they are said to be *identically distributed* if it holds that $\mathbb{P}(X_1 \in A') = \mathbb{P}(X_2 \in A')$ for all $A' \in \mathcal{A}'$. They are said to be *independent* if

$$\mathbb{P}(X_1 \in A_1', X_2 \in A_2') = \mathbb{P}(X_1 \in A_1')\mathbb{P}(X_1 \in A_1')$$

for all $A_1', A_2' \in \mathcal{A}'$. Note that this definition can be canonically extended to the case of n random variables.

The kth moment $\mathbb{E}(X^k)$ of a real–valued random variable X with $\int_{-\infty}^{\infty} |x^k| dF_X(x) < \infty$ can be given by

$$\mathbb{E}(X^k) = \int_{\Omega} X^k(\omega)\mathbb{P}(d\omega) = \int_{-\infty}^{\infty} x^k dF_X(x), \ k \in \mathbb{N},$$

where F_X is the cdf of X. More specifically, $\mathbb{E}(X)$ is said to be the *expectation* or the *mean* of X. Moreover, $Var X$ defined by

$$Var X = \mathbb{E}(X - \mathbb{E}X)^2 = \mathbb{E}X^2 - (\mathbb{E}X)^2$$

is called the *variance* of X.

In the following some specific distributions F_X for a real–valued random variable X are introduced. A (real–valued) random variable X is called *Poisson distributed* with $\mathbb{E}(X) = \lambda$ if $X : \Omega \to \mathbb{N}_0$ and

$$\mathbb{P}(X = k) = \exp{(-\lambda)}\frac{\lambda^k}{k!},$$

for $k \in \mathbb{N}_0$.

We call a (real–valued) random variable $X : \Omega \to [a, b] \subset \mathbb{R}$ *uniformly distributed* on the interval $[a, b]$ if its cdf F_X is given by

$$F_X(x) = \int_a^x \frac{1}{b - a} dt = \frac{x - a}{b - a}$$

for $x \in [a, b]$. Note that $\mathbb{E}(X) = (b + a)/2$.

A (real–valued) random variable $X : \Omega \to [0, \infty)$ with $\mathbb{E} = \lambda^{-1}$ is called *exponentially distributed* if its cdf F_X can be given by

$$F_X(x) = \int_0^x \lambda \exp^{-\lambda t} dt = 1 - \exp^{-\lambda x},$$

for $x \in [0, \infty)$.

We call a (real–valued) random variable $X : \Omega \to \mathbb{R}_+ \cup \{0\}$ with mean $\alpha\beta$ and variance $\alpha\beta^2$, where $\alpha, \beta > 0$, γ–*distributed* if its pdf $f_X(x)$ can be given by

$$f_X(x) = \frac{\beta^{-\alpha}}{\Gamma(\alpha)} x^{\alpha-1} \exp^{-\frac{x}{\beta}},$$

for $x \in [0, \infty)$.

A (real–valued) random variable $X : \Omega \to \mathbb{R}$ with mean $\mu \in \mathbb{R}$ and variance $\sigma^2 > 0$ is called *normally distributed* if its cdf F_X is given by

$$F_X(x) = \frac{1}{(2\pi\sigma^2)^{1/2}} \int_{-\infty}^x \exp^{-\frac{1}{2}(\frac{t-\mu}{\sigma})^2} dt,$$

for $x \in \mathbb{R}$. If furthermore $\mu = 0$ and $\sigma = 1$ we call X *standard normal distributed*.

An important theorem that is connected to the normal distribution is the central limit theorem.

Theorem A.1 (*Central Limit Theorem*) *Let $\{X_i\}_{i \geq 1}$ be a sequence of independent and identically distributed (real–valued) random variables with (common) expectation μ and variance $\sigma^2 > 0$. Then, if $S_n = \sum_{i=1}^n X_i$ it holds that*

$$\frac{S_n - n\mu}{\sigma\sqrt{n}} \xrightarrow{d} Z,$$

where Z follows a standard normal distribution.

For a proof of this theorem see, for example, [19], pp. 216ff.

Appendix B

Zusammenfassung

B.1   Ziele

Diese Arbeit basiert auf Ergebnissen die im Rahmen zweier noch andauernder Forschungsprojekte des Instituts für Stochastik der Universität Ulm erzielt wurden. Innerhalb des ersten Projekts, das in Kooperation mit France Télècom R&D Division, Paris, durchgeführt wird, steht im Vordergrund die Modellierung und die Analyse von Netzwerkstrukturen, die im Bereich der Telekommunikation auftreten. Insbesondere werden Straßensysteme städtischer Regionen wie beispielsweise Paris, aber auch nationale Telekommunikationsnetzwerke analysiert, wobei hauptsächlich die (zufällige) geometrische Struktur, sowie damit verbundene Kostenfunktionale von Interesse sind. Ziele der gewählten Modellansätze sind daher z.B. Aussagen über mittlere Abstände der Kunden zum nächstliegenden Telekommunikationsknoten oder über damit verbundene Kapazitätsfunktionale zu erhalten.

Das zweite Projekt, das in Zusammenarbeit mit Kollegen aus der Abteilung Innere Medizin I der Universität Ulm, der Zentralen Einrichtung Elektronenmikroskopie der Universität Ulm, des Laboratory of Cell and Computational Biology der University of California, der Abteilung Physik der Universität Leipzig sowie des Departments of Biology der University of Pennsylvania durchgeführt wird, beschäftigt sich mit der Untersuchung von intrazellulären Strukturen in menschlichen Zellen. Insbesondere ist hierbei das Zytoskelett von Interesse, welches innerhalb der Zelle netzwerkartige Strukturen ausbildet und somit der Zelle einerseits Stabilität verleiht und andererseits eine Schlüsselrolle für deren Fortbewegung einnimmt. Modelle für solche Skelettstrukturen können Informationen über abhängige Größen wie beispielsweise die Zellelastizität liefern.

Obwohl die beiden zu Beginn beschriebenen Projekte auf den ersten Blick nicht sehr viel

Gemeinsamkeiten hinsichtlich der untersuchten Objekte aufweisen, wurde es im Verlauf
dieser Arbeit immer klarer, dass die zugrundeliegenden mathematischen Modelle, Verfahren und Techniken, die zur Beantwortung der entstehenden Fragen dienen, sehr ähn-
lich sind. In beiden Fällen können die erhobenen Daten durch Netzwerkstrukturen
dargestellt werden, entweder auf einer makroskopischen (Telekommunikation) oder einer
mikroskopischen bzw. nanoskopischen Skala (Zytoskelett). Die Verbindungen zwischen
zwei Ecken oder Knoten können relativ gut durch Liniensegmente beschrieben werden,
was zu polygonähnlichen Strukturen führt. Diese Strukturen haben in beiden Projek-
ten dazu geführt, dass zur Modellierung zufällige Mosaike verwendet werden, die als
zufällige Partition der Ebene mit nichtleeren und nichtüberlappenden Polygonen aufge-
fasst werden können. Aus einem solchen Modellansatz ergibt sich die Notwendigkeit
geeignete Modelle für zufällige Mosaike auszuwählen, die einerseits geeignet sind, die
Daten gut widerzuspiegeln und andererseits immer noch mathematisch handhabbar
sind, z.b. hinsichtlich von Mittelwertsformeln für einfache geometrische Charakteristiken. Zusätzlich ist es häufig von Interesse, nach der Bestimmung eines geeigneten
Mosaikmodells, gewisse weitere Modellcharakteristiken zu bestimmen. Dies können
beispielsweise der mittlere Umfang der Polygone sein, die das Mosaik bilden oder auch
der mittlere Abstand eines zufällig gewählten Punkts zum Mittelpunkt des Polygons
des Mosaiks, indem dieser enthalten ist.

Im Speziellen werden im ersten Teil dieser Arbeit effiziente Schätzer für Kostenfunktionale in hierarchischen Modellen, die auf zufälligen Punktprozessen und zufälligen
Mosaiken basieren, hergeleitet und angewandt. Die Ergebnisse solcher Schätzungen
können beispielsweise für Kostenberechnungen oder Risikoanalysen dienen. Um zu
solchen effizienten Schätzern zu gelangen werden Grundkenntnisse über Themenbereiche
innerhalb der stochastischen Geometrie benötigt. Desweiteren müssen Simulationsalgo-
rithmen für die typische Zelle von speziellen Arten von zufälligen Mosaiken entwickelt
werden. Typische Zelle bedeutet hierbei eine vollkommen zufällig ausgewählte Zelle aus
den entstehenden Zellen des Mosaiks.

Der zweite Teil dieser Arbeit beginnt mit einer Beschreibung der angewandten Verfahren
zur Bildsegmentierung. Insbesondere werden die Bilddaten der Filamentnetzwerke des
Zytoskeletts in Graphenstrukturen umgewandelt, die für spätere statistische Untersuchungen geeignet sind. An diese Graphenstrukturen wird im Folgenden ein Modell
eines zufälligen Mosaiks angepasst, wobei der Anpassungsalgorithmus auf dem Vergleich globaler Netzwerkcharakteristiken wie z.B. der Anzahl der Ecken (Knoten) oder
der Gesamtlänge der Kanten pro Flächeneinheit, zwischen der beobachteten Graphen-
struktur und den theoretischen Modellen beruht. Die angepassten Modelle können dann
wiederum zur Schätzung weiterer Größen wie beispielsweise der Maschenbreite genutzt
werden.

Zusammenfassend werden in der vorliegenden Arbeit einige Ergebnisse der beiden beschriebenen Projekte vorgestellt. Dadurch soll der Leser von der Universalität der vorgestellten Methoden der stochastischen Geometrie und der morphologischen Bildanalyse bezüglich ihrer Anwendbarkeit auf unterschiedlichste Gebiete der wissenschaftlichen und industriellen Forschung überzeugt werden. Methoden, die verwendet werden, um eine spezielle Frage für eine bestimmte Anwendung zu beantworten, können oft in nur leicht abgewandelter Form dazu dienen, ein Problem in einem vollkommen anderen Anwendungsgebiet zu lösen. Noch genauer sind die Ziele dieser Arbeit:

- Die Entwicklung und Beschreibung effizienter Algorithmen zur Simulation der typischen Zelle für verschiedene Mosaikmodelle. Zusätzlich zu der Tatsache, dass die Anwendung solcher Algorithmen wertvolle Informationen über die Charakteristiken der typischen Zelle selber bereit stellt, dienen sie auch als Grundlage für spätere effiziente Kostenanalysen.

- Eine solche effiziente Kostenanalyse für hierarchische Modelle durchzuführen. Die Ergebnisse können anschließend beispielsweise realistische Kostenberechnungen im Bereich der Telekommunikation ermöglichen oder Erkenntnisse liefern hinsichtlich biologischer Prozesse, die im Zusammenhang mit dem Transport von Vesikeln in intrazellulären Strukturen stehen.

- Einige Fragen hinsichtlich der Möglichkeiten für Softwaretests zu beantworten, wobei hier das besondere Augenmerk auf das Testen von Software mit zufälligen Ein- bzw. Ausgaben gerichtet ist.

- Einige nützliche Methoden und Verfahren der morphologischen Bildanlyse vorzustellen, die in der Lage sind gegebene Bilddaten im Hinblick auf nachgelagerte statistische Analysen vorzuverarbeiten.

- Statistische Analysen für zwei Beispiele aus dem Bereich der Zellbiologie durchzuführen. Hierbei ist neben einer Untersuchung grundlegender statistischer Kenngrößen hauptsächlich ein Anpassungsalgorithmus für Modelle zufälliger Mosaike von Interesse. Ein solcher Anpassungsalgorithmus ist natürlich nicht nur auf zellbiologische Daten, sondern beispielsweise auch auf städtische Infrastrukturdaten anwendbar.

- Durch alle obig genannten Ziele aufzuzeigen, daß die Methoden und Verfahren der stochastischen Geometrie und der morphologischen Bildanalyse sehr vielseitig sind im Sinne einer Anwendbarkeit auf unterschiedlichste Problemstellungen für Bilddaten deren Maßstäbe makro–, mikro– oder sogar nanoskopisch sein können.

B.2   Gliederung

Nach der Einleitung in Kapitel 1 werden einige grundlegende Konzepte der stochastischen Geometrie in Kapitel 2 erläutert. Ziel ist es dem Leser genügend Grundlagen
zur Verfügung zu stellen, um den Begriff der zufälligen Mosaike einführen zu können,
da diese in allen späteren Anwendungen Verwendung finden. In Abschnitt 2.1 werden
zufällige abgeschlossene Mengen definiert, während in Abschnitt 2.2 zufällige Punktprozesse eingeführt werden, sowohl für den markierten als auch für den unmarkierten
Fall.   In diesem Abschnitt werden ebenfalls Palmverteilungen für zufällige Punktprozesse sowie Neveus Austauschformel für solche Palmverteilungen besprochen, die
in den Kapiteln 3 und 4 ihre Anwendung finden. Weitere wichtige Modelle aus der
stochastischen Geometrie, die in den Kapiteln 3 und 4 benötigt werden, sind Boolesche
Modelle und darauf basierende modulierte Poissonprozesse. Sie werden daher in Abschnitt 2.3 eingeführt. In Abschnitt 2.4 werden schließlich deterministische und zufällige
Mosaike definiert. Verschiedene Beispiele werden vorgestellt, die im Verlauf dieser Arbeit von Bedeutung sind, unter anderem das Poisson–Voronoi Mosaik, das Poissonsche
Geradenmosaik, das Poisson–Delaunay Mosaik, Cox–Voronoi Mosaike, modulierte Mosaike, Superpositionen und Nestings. Es sei an dieser Stelle bemerkt, dass hauptsächlich
der planare Fall betrachtet wird, wobei allerdings meistens kanonische Erweiterungen
hinsichtlich höherer Dimensionen existieren.

In Kapitel 3 werden einige effiziente Algorithmen zur Simulation typischer Zellen für
verschiedene zufällige Mosaike beschrieben. Abgesehen von der Möglichkeit durch deren
Anwendung Informationen über Charakteristiken der typischen Zelle zu erhalten, werden diese Algorithmen auch in Kapitel 4 hinsichtlich der Herleitung von effizienten
Schätzern für Kostenfunktionale in hierarchischen Modellen von Bedeutung sein.   In
Abschnitt 3.1 werden einige allgemeine Aspekte solcher Algorithmen für die Simulation
der typischen Zelle eines zufälligen Mosaiks diskutiert. Begonnen wird mit Algorithmen
zur Simulation eines Poissonschen Punktprozesses, der die Grundlage für alle Simula-
tionsalgorithmen darstellt, die in diesem Kapitel betrachtet werden. Slivnyaks Theorem
wird erläutert, welches die Darstellung der Palmverteilung eines Poissonschen Punktprozesses mit Hilfe seiner (unbedingten) Verteilung und eines zusätzlichen (determinis-
tischen) Punkts im Nullpunkt ermöglicht. Basierend auf dem Simulationsalgorithmus
des Poissonschen Punktprozesses und auf Slivnyaks Theorem wird ein Simulationsalgorithmus für die typische Zelle eines Poisson–Voronoi Mosaiks hergeleitet. Dieser
Simulationsalgorithmus wird in Abschnitt 3.2 erweitert, um die Simulation der typischen Zelle eines Cox–Voronoi Mosaiks zu ermöglichen, wobei das Cox–Voronoi Mosaik durch einen stationären Coxprozess auf den Linien eines Poissonschen Geradenmosaiks induziert wird.   Einige Resultate numerischer Auswertungen bezüglich der
Charakteristiken der typischen Zelle wie der empirischen Verteilung der Fläche oder
des Umfangs werden gezeigt. In Abschnitt 3.3 wird ein Algorithmus zur Simulation der

typischen Zelle eines modulierten Poisson–Voronoi Mosaiks vorgestellt. Hier wird der Fall betrachtet, dass das Voronoi Mosaik auf einem Coxprozess basiert, dessen zufälliges Maß durch ein Boolesches Modell mit kreisförmigen Körnern eines fixen oder zumindest beschränkten Radiuses erzeugt wird. Numerische Resultate einiger Charakteristiken wie der Verteilungen der Fläche der typischen Zelle für spezielle Parameterkonfigurationen bilden den Abschluss dieses Kapitels.

Ziel von Kapitel 4 ist die Herleitung von effizienten Schätzern für Kostenfunktionale in zwei verschiedenen hierarchischen Modellen. Als Vorbereitung werden zunächst in Abschnitt 4.1 einige grundlegende Begriffe der Graphentheorie eingeführt. Insbesondere werden, nach der Definition eines Graphen, einige bekannte Algorithmen zur Berechnung kürzester Wege und deren Längen in gegebenen Graphen besprochen. Das erste hierarchische Modell, welches in Abschnitt 4.2 untersucht wird, basiert auf zwei Coxprozessen, deren zufällige Intensitätsmaße sich auf die Linien eines gemeinsamen Poissonschen Geradenmosaiks konzentrieren. Eine Charakteristik von besonderem Interesse ist die kürzeste Weglänge, d.h. der Abstand entlang der Linien zwischen einem Punkt niederer Ordnung und seines (im Euklidischen Sinne) nächsten Nachbars höherer Ordnung. Eine weitere, eng verwandte Größe ist die sogenannte Subscriber Line Länge. Hier sind die Punkte niederer Ordnung nicht entlang der Linien platziert sondern rein zufällig in der gesamten Ebene verteilt und werden anschließend auf das nächstliegende Liniensegment innerhalb derselben Voronoizelle bzgl. der Punkte höherer Ordnung projiziert. Die Subscriber Line Länge ist dann durch den Abstand des projizierten Punkts niederer Ordnung zu dem Punkt höherer Ordnung der besagte Voronoizelle gegeben. Effiziente Schätzer für sowohl die mittlere kürzeste Weglänge als auch die mittlere Subscriber Line Länge werden hergeleitet, wobei die Algorithmen zur Simulation der typischen Zelle aus Kapitel 3 und Neveus Austauschformel für Palmverteilungen (Abschnitt 2.2) Verwendung finden. Ergebnisse von Monte–Carlo Simulationen zeigen in Verbindung mit theoretischen Resultaten zur Skalierungsinvarianz innerhalb dieses Modells Möglichkeiten für eine Nutzung der Methoden zur Kostenberechnung auf. In Abschnitt 4.3 wird ein weiteres hierarchisches Modell erläutert, dessen Grundlage die in Abschnitt 2.3 eingeführten modulierten Poissonschen Punktprozesse darstellen. Im Fokus der Untersuchungen ist hierbei der mittlere Abstand eines Punkts niederer Ordnung zu seinem nächsten Nachbarpunkt höherer Ordnung. Zu diesem Zweck wird ein effizienter Schätzer hergeleitet, der auf den Algorithmen zur Simulation der typischen Zelle (Kapitel 3) sowie auf Neveus Austauschformel für Palmverteilungen (Abschnitt 2.2) basiert. Einige numerische Resultate und die Diskussion von Möglichkeiten zur Nutzung hinsichtlich einer effizienten Kostenanalyse stehen am Ende von Kapitel 4.

Das Thema von Kapitel 5 sind Methoden zum Testen von Software mit zufälliger Ein– oder Ausgabe. In diesem Zusammenhang stellen Tests unter Verwendung eines statistischen Orakels einen zentralen Baustein dar. Diese werden in Abschnitt 5.2 erläutert. Einige Beispiele von Tests für Implementationen der Algorithmen aus den Kapiteln 3

und 4 werden besprochen. Abschnitt 5.3 beschäftigt sich mit Tests, die ein statisti-
sches Orakel mit einem Verfahren kombinieren, das metamorphisches Testen genannt
wird. Für diese Art von Softwaretests werden ebenfalls einige Beispiele demonstriert,
die die Korrektheit der Algorithmen aus Kapitel 3 überprüfen. Eine dritte Klasse
an Tests wird in Abschnitt 5.4 betrachtet. Bei dieser Art von Tests wird zusätzlich
zu einem statistischen Orakel und einer metamorphischen Beziehung noch eine weitere,
gegebene (und möglichst bereits getestete) Implementierung betrachtet, der sogenannte
Goldstandard. Nach einer Einführung in diese Klasse von Tests werden einige Beispiele
für Anwendungen gegeben. Kapitel 5 endet mit einer Zusammenfasssung und einem
Vergleich der betrachteten Testmethoden.

Kapitel 6 stellt eine Einführung in grundlegende Begriffe und Methoden der morphologischen Bildanalyse dar. Diese werden in Kapitel 8 dazu verwendet gegebene Bilddaten aus der Elektronenmikroskopie vorzuverarbeiten, um eine nachgelagerte statistische Analyse sowie ein Anpassen eines geeigneten Mosaikmodells zu ermöglichen. In Abschnitt 6.1 werden (digitale) Gitter und unterschiedliche Arten von digitalen Bildern wie
z.B. Grauwert- oder RGB Bilder definiert. Zusätzlich werden Verfahren zur Bilderver-
besserung durch das Anwenden von Filtern besprochen. Abschnitt 6.2 widmet sich
der Beschreibung der Skelettierung durch morphologische Operatoren. Hierdurch wird
eine gegebene Struktur in eine neue Struktur transformiert, die eine Breite von einem
Pixel hat, aber Eigenschaften der alten Struktur, beispielsweise die Anzahl an ver-
bundenen Komponenten beibehält. Ein zur Skelettierung eng verwandtes Verfahren
ist die Wasserscheidentransformation, die in Abschnitt 6.3 behandelt wird. Im Detail wird der Algorithmus zur Wasserscheidentransformation mittels Immersion (Ein-
tauchen) erläutert. Der letzte Abschnitt dieses Kapitels betrachtet einige weitere morphologische Operatoren, die in Kapitel 8 zur Verbesserung der Segmentierungsergeb-
nisse angewandt werden. Darunter fallen Operatoren wie iterative Entbartung oder die
Fusionierung benachbarter Kreuzungspunkte.

In Kapitel 7 wird ein Verfahren zur Anpassung zufälliger Mosaike an gegebene Netzwerkstrukturen erläutert. Dieses Verfahren wird in Kapitel 8 auf Beispiele aus der
Zellbiologie angewandt, kann aber aufgrund seiner Allgemeinheit auch Zur Analyse an-
derer Arten von Netzwerkstrukturen wie beispielsweise städtische Infrastrukturdaten
verwendet werden. In Abschnitt 7.1 werden die verwendeten Kenngrößen der Ein-
gangsdaten sowie zugehörige Schätzer erklärt. Das eigentliche Anpassungsverfahren
einschließlich der Wahl der Abstandsfunktion und des optimalen Modells wird in Abschnitt 7.2 beschrieben.

Zwei Anwendungen für Bilddaten aus der Zellbiologie der in den Kapiteln 6 und 7
beschriebenen Techniken sind der Gegenstand von Kapitel 8. Im ersten Beispiel wird
eine statistische Analyse von Keratin Filamentstrukturen durchgeführt. Solche Filamentstrukturen befinden sich im Zytoskelett von Epithelzellen und spielen für die

Migration und Stabilität der Zelle eine entscheidende Rolle. Hauptziel der Untersuchung ist die Detektion und Beschreibung struktureller Veränderungen der Architektur des Keratinnetzwerks bei Hinzugabe einer tumorfördernden Substanz. Nach einer Erläuterung der Datengewinnung und –segmentierung wird als erstes eine statistische Analyse grundlegender Charakteristiken, wie beispielsweise Orientierung und Länge der Filamentsegmente vollzogen. Zufällige Mosaike werden unter Verwendung des in Kapitel 7 beschriebenen Verfahrens an die Netzwerkstrukturen angepasst. Die Ergebnisse der Modellanpassung zeigen, dass tatsächlich eine Umstrukturierung der Netzwerkstruktur beobachtet werden kann, die durch Hinzugabe einer tumorfördernden Substanz induziert wird. Desweiteren kann diese durch die angewandten mathematischen Verfahren und Modelle sowohl qualitativ als auch quantitativ beschrieben werden.

Im zweiten Anwendungsbeispiel aus dem Bereich der Zellbiologie, das der Gegenstand von Abschnitt 8.2 ist, werden Aktin Filamentnetzwerke betrachtet, die ebenso wie die Keratinnetzwerke, einen Teil des Zytoskeletts in Zellen bilden. Diese Aktinnetzwerke sind verantwortlich für die Regulierung der Zellelastizität wodurch natürlich auch die Zellmigration beeinflusst wird. Ziel der Untersuchungen ist es ein zufälliges Mosaik zu finden, das die grundlegenden Eigenschaften der Netzwerkstrukturen in realen Beispieldaten möglichst gut widerspiegelt. Anschließend werden, basierend auf diesem angepassten Mosaikmodell, Näherungen für den elastischen Schermodul berechnet, der ein gutes Maß für die Zellelastizität darstellt. In Abschnitt 8.2.1 wird der angewandte Bildsegmentierungsalgorithmus detailliert beschrieben, dessen Grundlage die in Abschnitt 6.3 erklärte morphologische Wasserscheidentransformation ist. Ergebnisse der Anpassung von einfachen Superpositionsmodellen sind in Abschnitt 8.2.2 zu finden. Das hieraus resultierende optimale Mosaikmodell wird in Abschnitt 8.2.3 dazu verwendet Schätzungen für den elastischen Schermodul zu erhalten. Diese Schätzungen werden anschliessend mit rein auf Konzentrationen basierenden Schätzungen verglichen. Im Verlauf dieser Untersuchungen wird deutlich, dass der Ansatz des Anpassens von zufälligen Mosaiken geeignet ist, um nützliche Informationen über verschiedenste Größen der biologischen Zelle zu erhalten.

In Kapitel 9 werden die Ergebnisse der vorliegenden Arbeit zusammengefasst. Des Weiteren wird ein Ausblick auf darüber hinausgehende Fragestellungen und Probleme gegeben. Speziell kommen Verfahren zur Simulation von typischen Zellen für weitere Arten von zufälligen Mosaiken wie beispielsweise des aggregierten Poisson–Voronoi Mosaiks zur Sprache. Es werden mögliche Erweiterungen der beschriebenen Methoden zur effizienten Schätzung von Kostenfunktionalen in hierarchischen Modellen diskutiert, z.B. im Hinblick auf die Schätzung von Verteilungen und Dichten. Eine kurze Betrachtung von dynamischen Modellierungsansätzen und von Kostenanalysen im Bereich der Zellbiologie bilden den Schlusspunkt dieses Kapitels.

Zur Abrundung werden im Appendix einige grundlegende mathematische Begriffe eingeführt. Die ersten beiden Abschnitte behandeln vornehmlich Themen aus der Mengen-theorie und der Topologie wie beispielsweise Mengen, metrische Räume und Dichtheit. In den letzten beiden Abschnitten stehen Konzepte aus der Maßtheorie und der Wahr-scheinlichkeitsrechnung im Vordergrund. Als Beispiele seien hier Maße, Verteilungen und der zentrale Grenzwertsatz genannt.

Bibliography

[1] Alberts, B., Bray, D., Lewis, J., Raff, M., Roberts, K., Watson, J. D. (1994) *Molecular Biology of the Cell*. Garland Publishing, New York.

[2] Ambartzumian, R. V. (2000) *Factorization Calculus and Geometric Probability*. Cambridge University Press, Cambridge.

[3] Ananthakrishnan, R., Guck, J., Käs, J. (2006) Cell mechanics: Recent advances with a theoretical perspective. *Recent Research and Development in Cell Biology* (in press).

[4] Baccelli, F., Klein, M., Lebourges, M., Zuyev, S. (1997) Stochastic geometry and architecture of communication networks. *Telecommunications systems* **7**, 209–227.

[5] Baccelli, F., Klein, M., Lebourges, M., Zuyev, S. (1996) Géométrie aléatoire et architecture de réseaux. *Annales des Télécommunications* **51**, 158–179.

[6] Batchelet, E. (1981) *Circular Statistics in Biology*. Academic Press, London.

[7] Bauer, H. (1992) *Maß und Integrationstheorie*. Second edition, de Gruyter, Berlin.

[8] Beil, M., Braxmeier, H., Fleischer, F., Schmidt, V., Walther, P. (2005) Quantitative analysis of keratin filament networks in scanning electron microscopy images of cancer cells. *Journal of Microscopy* **220**, 84–95.

[9] Beil, M., Eckel, S., Fleischer, F., Schmidt, H., Schmidt, V., Walther, P. (2006) Fitting of random tessellation models to cytoskeleton network data. *Journal of Theoretical Biology* **241**, 62–72.

[10] Beil, M., Fleischer, F., Paschke, S., Schmidt, V. (2005) Statistical analysis of 3D centromeric heterochromatin structure in interphase nuclei, *Journal of Microscopy* **217**, 60–68.

[11] Beil, M., Micoulet, A., von Wichert, G., Paschke, S., Walther, P., Omary, M. B., van Veldhoven, P., Gern, U., Wolff–Hieber, E., Eggermann, J., Waltenberger, J., Adler, G., Spatz, J., Seufferlein, T. (2003) Sphingosylphosphorylcholine regulates keratin network architecture and visco–elastic properties of human cancer cells. *Nature Cell Biology* **5**, 803–811.

[12] Benes, V., Rataj, J. (2004) *Stochastic Geometry: Selected Topics.* Kluwer, Dordrecht.

[13] Billingsley, P. (1986) *Probability and Measure.* Second edition, John Wiley & Sons, Inc., New York.

[14] Binder, R. V. (2000) *Testing Object–Oriented Systems: Models, Patterns, and Tools.* Addison–Wesley, Reading.

[15] Błaszczyszyn, B., Schott, R. (2005) Approximation of functionals of some modulated Poisson–Voronoi tessellations with applications to modelling of communication networks. *Japan Journal of Industrial and Applied Mathematics* **22**, 179–204.

[16] Błaszczyszyn, B., Schott, R. (2003) Approximate decomposition of some modulated Poisson–Voronoi tessellations. *Advances in Applied Probability* **35**, 847–862.

[17] Blum, M., Kannan, S. (1989) Designing programs that check their work. In *Proceedings of the 31st Annual ACM Symposium on Theory of Computing (STOC'89)*, 86–97. ACM Press, New York.

[18] Blum, M., Luby, M., Rubinfeld, R. (1993). Selftesting/correcting with applications to numerical problems. *Journal of Computer and System Sciences* **47**(3), 549–595.

[19] Casella, C., Berger, R. L. (2002) *Statistical Inference.* Wadsworth, Duxbury.

[20] Chen, T. Y., Kuo, F.–C., Tse, T. H., Zhi Q. Z. (2003) Metamorphic testing and beyond. In *Proceedings of the International Workshop on Software Technology and Engineering Practice (STEP)*, 94–100. IEEE Computer Society Press, Los Alamitos, California.

[21] Chung, K.L. (2000) *A Course in Probability Theory.* Third edition, Academic Press, New York.

[22] Daley, D.J., Vere–Jones, D. (1988) *An Introduction to the Theory of Point Processes.* Springer, New York.

[23] Diestel, R. (2000) *Graphentheorie*. Springer, New York.

[24] Fleischer, F. (2002) *Routing in Telecommunication Access Networks. Path Optimization for the Stochastic Subscriber Line Model*. Diploma thesis, Ulm University.

[25] Fleischer, F., Eckel, S., Schmid, I., Schmidt, V., Kazda, M. (2006) Point process modelling of root distribution in pure stands of Fagus sylvatica and Picea abies. *Canadian Journal of Forest Research* **36**(1), 227–237.

[26] Fleischer, F., Beil, M., Ananthakrishnan, R., Eckel, S., Schmidt, H., Käs, J., Svitkina, T., Schmidt, V. (2007) Actin network architecture and elasticity in lamellipodia of melanoma cells. *New Journal of Physics* **9** 420.

[27] Fleischer, F., Beil, M., Kazda, M., Schmidt, V. (2006) Analysis of spatial point patterns in microscopic and macroscopic biological image data. In A. Baddeley, P. Gregori, J. Mateu, R. Stoica, D. Stoyan (eds.) *Case studies in spatial point processes models, Lecture Notes in Statistics*, 235–260. Springer, Berlin.

[28] Fleischer, F., Gloaguen, C., Schmidt, H., Schmidt, V., Schweiggert, F. (2008) Simulation of typical modulated Poisson–Voronoi cells and their application to telecommunication network modelling. Preprint (submitted for publication).

[29] Foss, S. G., Zujev, S. (1996) On a Voronoi aggregate process related to a bivariate Poisson process. *Advances in Applied Probability* **28**, 956–981.

[30] Gerig, G., Kübler, O., Kikinis, R., Jolesz, F. A. (1992) Nonlinear anisotropic filtering of MRI data. *IEEE Transactions on Medical Imaging* **11**(2), 221–232.

[31] Gloaguen, C., Fleischer, F., Schmidt, H., Schmidt, V. (2006) Simulation of typical Cox–Voronoi cells, with a special regard to implementation tests. *Mathematical Methods of Operations Research* **62**, 357–373.

[32] Gloaguen, C., Fleischer, F., Schmidt, H., Schmidt, V. (2006) Fitting of stochastic telecommunication network models, via distance measures and Monte–Carlo tests. *Telecommunications Systems* **31**, 353–377.

[33] Gloaguen, C., Fleischer, F., Schmidt, H., Schmidt, V. (2006) Modelling and simulation of telecommunication networks: analysis of mean shortest path lengths. In R. Lechnerová, I. Saxl, V. Beneš (eds.) *Proceedings of the 6th International Conference on Stereology, Spatial Statistics and Stochastic Geometry*, 25–36. Union of Czech Mathematicians and Physicists, Prague.

[34] Gloaguen, C., Fleischer, F., Schmidt, H., Schmidt, V. (2008) Analysis of short-
     est paths and subscriber line lengths in telecommunication access networks.
     Networks and Spatial Economics (to appear).

[35] Gonzalez, R., Woods, R. (2002) *Digital Image Processing*. Second edition,
     Prentice Hall, New York.

[36] Halmos, P. R. (1950) *Measure Theory*. Springer, New York.

[37] Harding, E. F., Kendall, W. S. (1974) *Stochastic Geometry*. John Wiley &
     Sons, New York.

[38] Hörner, S. (2004) *Modelling of Telecommunication Networks. Simulation of
     Random Tessellations and their Statistical Fitting to Real Infrastructure Data*.
     Diploma thesis, Ulm University.

[39] Hoffman, D. (1999) Heuristic test oracles. *Software Testing and Quality En-
     gineering* **1**, 29–32.

[40] Hornischer, A. (2006) *Statistical Analysis of Multi–Type Nestings and Asymp-
     totic Normality of Functionals of Tessellations*. Diploma thesis, Ulm Univer-
     sity.

[41] Jähne, B. (2002) *Digital Image Processing*. Springer, New York.

[42] Jungnickel, D. (1997). *Graphs, Networks and Algorithms*. Springer, Berlin.

[43] Käs, J., Strey, H., Tang, J. X., Finger, D., Ezzell, R., Sackmann, E., Jan-
     mey, P.A. (1996) F–actin, a model polymer for semiflexible chains in dilute,
     semidilute, and liquid crystalline solutions. *Biophysical Journal* **70**, 609–625.

[44] Kallenberg, O. (1986) *Random Measures*. Forth edition, Akademie–Verlag,
     Berlin and Academic Press, London.

[45] Karr, A.F. (1993) *Probability*. Springer, Berlin.

[46] König, D., Schmidt, V. (1992) *Zufällige Punktprozesse*. B.G. Teubner,
     Stuttgart.

[47] Last, G., Brandt, A. (1980) *Marked Point Processes on the Real Line*.
     Springer, New York.

[48] Levene, H. (1960) Robust tests for the equality of variances. In: I. Olkin
     (ed.) *Contributions to Probability and Statistics*, 278–292. Stanford University
     Press, Palo Alto.

[49] Lodish, H., Berk, A., Matsudaira, Chris, P., Kaiser, A., Krieger, M., Scott, M. P., Zipursky, L. , Darnell, J. (2003) *Molecular Cell Biology*. Fifth edition. W. H. Freeman, New York.

[50] Lück, S. (2006) *A Stochastic Model for the Reorganization of the Keratin Cytoskeleton*. Diploma thesis, Ulm University.

[51] MacKintosh, F., Käs, J., Janmey, P. (1995) Elasticity of semiflexible biopolymer networks. *Physical Review Letters* **75**, 4425–4428.

[52] Maier, R. (2003) *Iterated Random Tessellations – With Application to Spatial Modelling of Telecommunication Networks*. Dissertation, Ulm University.

[53] Maier, R., Mayer, J., Schmidt, V. (2002) Distributional properties of the typical cell of stationary iterated tessellations. *Mathematical Methods of Operations Research* **59**, 287–302.

[54] Maier, R., Schmidt, V. (2003) Stationary iterated tessellations. *Advances in Applied Probability* **35**, 337–353.

[55] Matheron, G. (1975) *Random Sets and Integral Geometry*. John Wiley & Sons, New York.

[56] Matheron, G. (1967) *Elements pour une Theorie des Mileux Poreux*. Masson, Paris.

[57] Mattfeldt, T., Eckel, S., Fleischer, F., Schmidt, V. (2006) Statistical analysis of reduced pair correlation functions of capillaries in the prostate gland. *Journal of Microscopy* **223**, 107–119.

[58] Mattfeldt, T., Eckel, S., Fleischer, F., Schmidt, V. (2007) Statistical modelling of the geometry of prostatic capillaries on the basis of stationary Strauss hardcore processes. *Journal of Microscopy* **228**, 272–281.

[59] Mattfeldt, T., Fleischer, F. (2006) Computer–intensive methods for statistical interference from stereological data. In R. Lechnerová, I. Saxl, V. Beneš (eds.) *Proceedings of the 6th International Conference on Stereology, Spatial Statistics and Stochastic Geometry*, 349–360. Union of Czech Mathematicians and Physicists, Prague.

[60] Mattfeldt, T., Fleischer, F. (2005) Bootstrap methods for statistical inference from stereological estimates of volume fraction. *Journal of Microscopy* **218**, 160–170.

[61] Matthes, K., Kerstan, J., Mecke, J. (1978) *Infinitely Divisable Point Processes*. Second edition, John Wiley & Sons, Chichester.

[62] Mayer, J. (2003) *On Quality Improvement of Scientific Software: Theory, Methods, and Application in the GeoStoch Development*. Dissertation, Ulm University.

[63] Mayer, J., Guderlei, R. (2007) *An empirical study on the selection of good metamorphic relations*. Accepted by *COMPSAC 2006*

[64] Mayer, J., Guderlei, R. (2004) Test oracles using statistical methods. *Lecture Notes in Informatics* **58**, 85–95.

[65] Mayer, J., Schmidt, V., Schweiggert, F. (2004) A unified simulation framework for spatial stochastic models. *Simulation Modelling Practice and Theory* **12**, 307–326.

[66] Mardia, K. V., Jupp, P. E. (2000) *Directional Statistics*. John Wiley & Sons, Chichester.

[67] Mecke, J. (1984) Parametric representation of mean values for stationary random mosaics. *Math. Operationsforsch. Stat. Ser. Statist.* **15**, 437–442.

[68] Mecke, J., Schneider, R. G., Stoyan, D., Weil, W. R. R. (1990) *Stochastische Geometrie*. Birkhäuser, Basel.

[69] Molchanov, I. (2005) *Theory of Random Closed Sets*. Springer, London.

[70] Molchanov, I. (1997) *Statistics of the Boolean Model for Practitioners and Mathematicians*. John Wiley & Sons, Chichester.

[71] Møller, J. (1994) *Lectures on Random Voronoi Tessellations*. Lecture Notes in Statistics **87**, Springer, New York.

[72] Møller, J. (1989) Random tessellations in $\mathbb{R}^d$. *Advances in Applied Probability* **21**, 37–73.

[73] Morse, D. C. (1998) Viscoelasticity of concentrated isotropic solutions of semi-flexible polymers: 1. Model and stress tensors. *Macromolecules* **31**, 7030–7043.

[74] Müller, E. (2003) *Statistical Analysis and Approximation Formulae for Expected Means of Shortest Path Lengths in Telecommunication Networks*. Diploma thesis, Ulm University.

[75] Neveu, J. (1976) Sur les mesures de Palm de deux processus ponctuels sta-
tionnaires. *Zeitschrift für Wahrscheinlichkeitstheorie und verwandte Gebiete*
34, 199–203.

[76] Nogales E., Wang H. W. (2006) Structural intermediates in microtubule assembly and disassembly: how and why? *Current Opinion in Cell Biology*
18(2), 179–184.

[77] Ohser, J., Mücklich, F. (2000) *Statistical Analysis of Microstructures in Ma-
terials Science*. John Wiley & Sons, Chichester.

[78] Okabe A., Boots B., Sugihara K., Chiu S. N. (2000) *Spatial Tessellations*.
Second edition, John Wiley & Sons, Chichester.

[79] Omary M. B., Coulombe P. A., McLean W. H. (2004) Intermediate filament proteins and their associated diseases. *New England Journal of Medicine*
351(20), 2087–2100.

[80] Park, S., Koch, D., Cardenas, R., Käs, J., Shih, C. K. (2005) Cell motility
and local viscoelasticity of fibroblasts. *Biophysical Journal* **89**, 4330–4342.

[81] Perona, P., Malik, J. (1990) Scale-space and edge detection using anisotropic
diffusion. *IEEE Trans. on Pattern Analysis and Machine Intelligence* **12**(7),
629–639.

[82] Pollard T. D., Borisy G. G. (2003) Cellular motility driven by assembly and
disassembly of actin filaments. *Cell* **112**, 453–465.

[83] Ponti, A., Machacek, M., Gupton, S. L., Waterman-Storer, C. M., Danuser, G.
(2004) Two distinct actin networks drive the protrusion of migrating cells.
Science **305**(5691), 1782–1786.

[84] Posch, K. (2005) *Modulated Poisson–Voronoi Tessellations. Simulation of the
Typical Cell and Applications to Telecommunication Networks*. Diploma thesis, Ulm University.

[85] Press, W.H., Flannery, B.P., Teukolsky, S.A., Vetterling, W.T. (1986) *Numerical Recipes*. Cambridge University Press, Cambridge.

[86] Quine, M. P., Watson, D. F. (1984) Radial generation of n–dimensional Poisson processes. *Journal of Applied Probability* **21**, 548–557.

[87] Roerndink, J. B. T. M., Meijster, A. (2001) The watershed transform: Definitions, algorithms and parallelization strategies. *Fundamenta Informaticae* **41**,
187–228.

[88] Rösch, M. (2005) *Cost Analysis for Telecommunication Networks. Mean Value Formulae for Subscriber Line Lengths.* Diploma thesis, Ulm University.

[89] Royden, H. L. (1988) *Real Analysis.* Third edition, Macmillan, New York.

[90] Santaló, L. (1984) Mixed random mosaics. *Mathematische Nachrichten* **117**, 129–133.

[91] Santaló, L. (1976) *Integral Geometry and Geometric Probability.* Addison–Wesley, Reading.

[92] Schmidt, H. (2006) *Assymptotic Analysis of Stationary Random Tessellations with Applications to Network Modelling.* Dissertation, Ulm University.

[93] Schmidt, V. (2007) *Räumliche Statistik.* Lecture Notes, Ulm University.

[94] Schneider, R., Weil, W. (2000) *Stochastische Geometrie.* Teubner, Stuttgart.

[95] Schneider, R., Weil, W. (1992) *Integralgeometrie.* Teubner, Stuttgart.

[96] Serra, J.P. (1988) *Image Analysis and Mathematical Morphology.* Academic Press, London.

[97] Shao, D., Forge, A., Munro, P. M., Bailly, M. (2006) Arp2/3 complex–mediated actin polymerisation occurs on specific pre–existing networks in cells and requires spatial restriction to sustain lamellipod extension. *Cell Motility Cytoskeleton* Apr. 17; [Epub ahead of print].

[98] Slivnyak, I. M. (1962) Some properties of stationary flows of homogeneous random events. *Teor. Veruyat. Primen.* **7**, 347–352. (Translation in *Theory Probab. Appl.* **7**, 336–341).

[99] Soille, P. (2003) *Morphological Image Analysis.* Springer, Berlin.

[100] Stoyan, D., Kendall, W.S., Mecke, J. (1995) *Stochastic Geometry and its Applications.* Second edition, John Wiley & Sons, Chichester.

[101] Stoyan, D., Stoyan, H. (1995) *Fractals, Random Shapes and Point Fields.* John Wiley & Sons, Chichester.

[102] Svitkina, T. M., Borisy, G. G. (1998) Correlative light and electron microscopy of the cytoskeleton of cultured cells. *Methods Enzymology* **298**, 570–592.

[103] Svitkina, T. M., Bulanova, E. A., Chaga, O. Y., Vignjevic, D. M., Kojima, S., Vasiliev, J. M., Borisy, G. G. (2003) Mechanism of filopodia initiation by reorganization of a dendritic network. *Journal of Cell Biology* **160**, 409–421.

[104] Tang, J. X., Janmey, P. A. (1996) The polyelectrolyte nature of F–actin and the mechanism of actin bundle formation. *Journal of Bioligical Chemistry* **271**, 8556–8563.

[105] Tchoumatchenko, K., Zuyev, S. (2001) Aggregate and fractal tessellations. *Probability Theory Related Fields* **121**, 198–218.

[106] Thiedmann, R. (2006) *Optimization Methods for Statistical Fitting of Network Models and Applications to Intensity Maps.* Diploma thesis, Ulm University.

[107] Upowsky, A. (2006) *Stochastic Modelling and Cost Analysis of Nationwide Telecommunication Networks.* Diploma thesis, Ulm University.

[108] Van Lieshout, M. N. M. (2000) *Markov Point Processes and Their Applications.* Imperial College Press, London.

[109] Vincent, L. (1993) Morphological grayscale reconstruction in image analysis: Applications and efficient algorithms. *IEEE Trans. Image Processing* **2**(2), 176–201.

[110] Vincent, L., Soille, P. (1991) Watersheds in digital spaces: An efficient algorithm based on immersion simulations. *IEEE Trans. Patt. Anal. Mach. Intell.* **13**(6).

[111] Wackernagel, H. (1998) *Multivariate Geostatistics.* Second edition, Springer, Berlin.

[112] Wauer, M. (2006) *Aggregated Poisson–Voronoi Tessellations. Simulation of the Typical Cell and Applications to Telecommunication Network Modelling.* Diploma thesis, Ulm University.

[113] Wendel, J. G. (1962) A problem in geometric probability. *Mathematica Scandinavia* **11**, 109–111.

[114] Wilcoxon, F. (1945) Individual Comparisons by Ranking Methods. *Biometrics* **1**, 80–83.

[115] Wolfmüller, D. (2006) *Cost Analysis for Telecommunication Networks. Least Nodes Paths and Capacity Loads.* Diploma thesis, Ulm University.

[116] Zigmond, S.H. (1993) Recent quantitative studies of actin filament turnover during cell locomotion. *Cell Motility Cytoskeleton* **25**, 309–316.

www.ingramcontent.com/pod-product-compliance
Lightning Source LLC
LaVergne TN
LVHW022310060326
832902LV00020B/3375